OTHER TITLES OF INTEREST FROM ST. LUCIE PRESS

Supply Chain Management: The Basics and Beyond

The Constraints Management Handbook

Total Productivity Management: A Systemic and Quantitative Approach to Compete in Quality, Price, and Time

Step-by-Step QFD: Customer-Driven Product Design

Introduction to the Theory of Constraints

Systematic Innovation: An Introduction to TRIZ

Reengineering Performance Management: Breakthroughs in Achieving Strategy Through People

Macrologistics Management

Project Management in the Fast Lane: Applying the Theory of Constraints

EU Directive Handbook: Understanding the European Union Compliance Process and What It Means to You

Total Quality in Information Systems and Technology

Fundamentals of Industrial Quality Control

For more information about these titles call, fax or write:

St. Lucie Press
2000 Corporate Blvd., N.W.
Boca Raton, FL 33431-9868

TEL (561) 994-0555 • (800) 272-7737
FAX (800) 374-3401
E-MAIL information@slpress.com
WEB SITE http://www.slpress.com

$S{^t_L}$

Applying
MANUFACTURING EXECUTION SYSTEMS

Michael McClellan

The St. Lucie Press/APICS Series on Resource Management

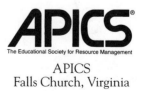

S_L^t

St. Lucie Press
Boca Raton, Florida

The Educational Society for Resource Management

APICS
Falls Church, Virginia

S^t_L

The Educational Society for Resource Management

St. Lucie Press
2000 Corporate Blvd., N.W.
Boca Raton, Florida 33431-9868

APICS
500 West Annandale Road
Falls Church, Virginia 22046-4274

This book is dedicated to my daughters,
Brenda and Debbie.

Contents

Preface

M anufacturing execution systems (commonly referred to as MES) are rapidly taking their place in the world of manufacturing. Unlike other manufacturing computer systems, such as Manufacturing Resources Planning (MRPII), that are quite well defined as to functionality and application, MES is specific only in concept.

This book defines what a manufacturing execution system is and explains how this concept can and is being applied within manufacturing companies. MES is not new. Since the early 1980s, a number of companies have used large and small computer systems to assist production operations. At that time, the general buzzword describing these applications was computer-integrated manufacturing (CIM), a term that has fallen into disuse primarily due to application disappointments. Although CIM and MES are very similar concepts, there has been significant development during the past few years: MES has taken on broader and, in some cases, totally different meaning, providing the infrastructure that CIM never had. Instead of custom designing software programs on a contract-by-contract basis, MES is based much more on standard reusable application software. The results include lower costs and shorter implementation time. Most important, the system is more likely to work as promised.

Another important change has been the general acceptance of three levels of control within manufacturing: the planning level (Material Requirements Planning [MRP], Enterprise Resources Planning [ERP], etc.) that defines what is to be manufactured within a given time period, the execution level (MES) that takes the planning output and executes the plan on a near real-time/on-line basis, and the control level (programmable logic controller [PLC]) that operates the equipment in the plant. As these layers have become better defined and accepted, the functionality within each layer has become more standard. The connections, both logical and physical, are better known and understood. As technological development continues, particularly in equipment, overlap between the layers will increase. But that overlap will most likely be based on extensions from one layer to another, not elimination or obsolescence of functionality. This book addresses the execution layer by providing a definition and understanding of what this layer includes—what the parts are and how they can work together. Why the execution layer is separate and different from the planning and/ or control layers will also be explained.

This book supports all types of manufacturing from discrete item production to process flow production. The concepts are applicable in all production facilities where a number of variables (simple or complex) need to be considered to optimize production by effectively using the available resources of people, inventory, and equipment. The idea of agile manufacturing using MES technology is discussed.

Although MES is usually not equipment or operating system specific, Chapter 7 addresses some issues regarding computers and communication equipment as well as data entry methods and equipment.

An important part of the MES discussion is the evolutionary nature of an MES system. A full MES application can and will include many functions within a company. It is unlikely that the initial MES implementation will include all of the possible functionality on the first day of start-up. Initial implementation will likely suggest changes—the inclusion of more and better ideas—that could not have been anticipated earlier. Allowance for their later implementation should be considered. This makes MES projects excellent vehicles for continuous improvement, a great tool for production management.

Manufacturing execution systems are computer system tools designed and built to help manufacturing management—from top management to the plant floor—do their job of accomplishing production most effectively. As the name implies, MES is more than a planning tool like MRP. MES is an on-line extension of MRP with emphasis on execution that includes:

- Making products
- Turning machines on and off
- Measuring parts
- Changing order priorities
- Setting and reading measuring controls
- Scheduling and rescheduling machines
- Assigning and reassigning inventory
- Moving inventory to and from workstations
- Assigning and reassigning personnel
- Managing the production process
- Setting alarms for out-of-process conditions

This book has been written to be a guide to those who have the primary responsibility of improving their companies' manufacturing process. It is about manufacturing and how to tie the many variables of manufacturing into a better production system through the use of an integrated computer system. It is not about bits and bytes, computers, or operating systems. The manufacturing professional and the information technology professional will find this book informative and helpful in applying MES as an effective tool to manage current events in order to most effectively accomplish the production plan.

Acknowledgments

Writing this book has required the appreciated help of a number of people. A special thank-you is extended to the companies that have provided application examples, the Manufacturing Execution Systems Association members who have given their time to provide review and comment on the book material, and the many equipment and software suppliers that have provided information on their products. Thanks also to Marge Hoye, Ph.D. for her organizing and editing effort and to Thom Porterfield for his direction and assistance on illustrations.

About the Author

Michael McClellan is president of MES Solutions Incorporated, a company specializing in consulting services aiding clients in the design and implementation of manufacturing execution systems. Prior to his current position, Mr. McClellan was president of a major supplier of material management and control systems. Before that, he was a founder and president of Integrated Production Systems, a company that pioneered the development and implementation of computer systems for production execution. While at Integrated Production Systems, he published two papers and numerous articles on the subject of applying computer systems in manufacturing operations and managed the company that developed and installed systems for major clients.

Prior to forming Integrated Production Systems, Mr. McClellan held officer-level management positions in companies providing equipment and control systems for production and material management.

In addition to his work experience, Mr. McClellan is a member of the Manufacturing Execution Systems Association (MESA International), American Production and Inventory Control Society (CPIM), and the Society of Manufacturing Engineers. He holds one patent.

Mr. McClellan is a fervent believer in MES as a significant production tool that is barely past the infancy stage. He is interested in any comments and ideas regarding MES and can be reached by telephone at (541) 548-6690, by fax at (541) 548-6674, or by mail at P.O. Box 2148, Terrebonne, Oregon 97760.

About APICS

A PICS, The Educational Society for Resource Management, is an international, not-for-profit organization offering a full range of programs and materials focusing on individual and organizational education, standards of excellence, and integrated resource management topics. These resources, developed under the direction of integrated resource management experts, are available at local, regional, and national levels. Since 1957, hundreds of thousands of professionals have relied on APICS as a source for educational products and services.

- **APICS Certification Programs**—APICS offers two internationally recognized certification programs, Certified in Production and Inventory Management (CPIM) and Certified in Integrated Resource Management (CIRM), known around the world as standards of professional competence in business and manufacturing.
- *APICS Educational Materials Catalog*—This catalog contains books, courseware, proceedings, reprints, training materials, and videos developed by industry experts and available to members at a discount.
- *APICS—The Performance Advantage*—This monthly, four-color magazine addresses the educational and resource management needs of manufacturing professionals.
- *APICS Business Outlook Index*—Designed to take economic analysis a step beyond current surveys, the index is a monthly manufacturing-based survey report based on confidential production, sales, and inventory data from APICS-related companies.

- **Chapters**—APICS' more than 270 chapters provide leadership, learning, and networking opportunities at the local level.
- **Educational Opportunities**—Held around the country, APICS' International Conference and Exhibition, workshops, and symposia offer you numerous opportunities to learn from your peers and management experts.
- **Employment Referral Program**—A cost-effective way to reach a targeted network of resource management professionals, this program pairs qualified job candidates with interested companies.
- **SIGs**—These member groups develop specialized educational programs and resources for seven specific industry and interest areas.
- **Web Site**—The APICS web site at http://www.apics.org enables you to explore the wide range of information available on APICS' membership, certification, and educational offerings.
- **Member Services**—Members enjoy a dedicated inquiry service, insurance, a retirement plan, and more.

For more information on APICS programs, services, or membership, call APICS Customer Service at (800) 444-2742 or (703) 237-8344 or visit http://www.apics.org on the World Wide Web.

Applying
MANUFACTURING EXECUTION SYSTEMS

ERP

MES

SPC
Time + Attend.
Recivng,
Inspecton

Control
Sys

What MES Is and Why It's a Hot Topic

Manufacturing execution systems have evolved to fill the communication gap between the manufacturing planning system (MRP, MRPII, ERP, etc.) and the control systems used to run equipment on the plant floor. Previously, these were the tools production management was given, along with a lot of printed paper and many disparate pieces of information such as statistical process control (SPC), time and attendance, receiving reports, inspection reports, etc., to accomplish the production task. It was up to production to determine how to tie all the information together. There were a few problems with this arrangement:

- The data was usually late.
- The data was rarely current or reliably correct.
- The information was voluminous and extremely difficult to assimilate.
- The information was usually based on another department's idea of what was important.

Computer software systems were developed to help production managers better use this information to **execute** the manufacturing plan; hence the name manufacturing execution systems. Once the name stuck, many companies offered solutions for specific narrow areas in manufacturing and called them manufacturing execution systems. SPC has been called an MES. Tracking work orders through production routing steps has been called MES. Data collection systems

1

have been called MES. There are many more examples that follow this same line, leaving most users confused and asking, "What is an MES?"

The most inclusive and specific definition of MES might be the following: **A manufacturing execution system (MES) is an on-line *integrated* computerized system that is the accumulation of the methods and tools used to accomplish production.** The purpose of this book is to provide an explanation of how this accumulated functionality can go beyond individual point solutions to an *integrated* system that is significantly more effective as a management tool.

The MES is primarily a formalization of production methods and procedures into an integrated computer system that presents data in a more useful and systematic form. If your company accomplishes production, there must be some kind of system in place to make it happen. In most companies, existing systems are made up of many non-integrated components and, in some cases, unidentifiable parts. In any case, some system exists. The definition differentiates a non-integrated system (the likely current system) from the on-line integrated capabilities of an MES. This may seem to be a distinction without a difference, but the idea is that MES is the step of **integrating** all of the activities that are not in the planning layer or in the device control layer as components or possible components of a proactive integrated on-line system providing a synergistic process that is greater than the sum of the parts.

The Manufacturing Execution Systems Association (MESA International) has conducted a study of user companies and offers the following list of benefits from using a computer-driven MES:

- Reduces manufacturing cycle time
- Reduces or eliminates data entry time
- Reduces work-in-process
- Reduces or eliminates paperwork between shifts
- Reduces lead times
- Improves product quality
- Eliminates lost paperwork
- Empowers plant operations people

- Improves the planning process
- Improves customer service

Other than a more coherent presentation of production information, an MES will add little to your production methods. Implementing SPC will not change your manufacturing process, but if used as a part of your production system, your decision-making information will be improved. Knowing what resources are available will not increase output, but the information may help to better establish priorities. The time and attendance system will not change production output, but knowing specific skills are available might allow more realistic planning. Knowing the current work center loadings will not change rates, but current knowledge allows you to make changes, find an order, or estimate completion. On-line inventory quantity and location information has no impact unless you need to know about a specific item when rescheduling work orders or quantities. Scheduling work into a work center down for maintenance will not have the same effect if your system advised you the resource was not available or automatically scheduled the work into an alternate resource. The primary advantage of the MES is the ability to integrate more accurate and current information into the decision-making process.

Another advantage is that an MES can be proactive, causing events to occur or tasks to be completed according to the plant's operating methods or plan and without human intervention. An example is the automatic download of a CAD program to a work center according to the workstation schedule. Another is the automatic movement of a specific item of inventory to a workstation following the part routing and order schedule.

MESA International has prepared a list of activity descriptions of the various areas of production management that would be included in a full MES implementation. They are as follows:

- **Resource Allocation and Status**—Manages resources, including machines, tools, labor skills, materials, and other equipment and other entities such as documents that must be available for work to start at an operation.

- **Operations/Detail Scheduling**—Provides sequencing based on priorities, attributes, characteristics, and/or recipes associated with specific production units at an operation.
- **Dispatching Production Units**—Manages the flow of production units in the form of jobs, orders, batches, lots, and work orders.
- **Document Control**—Controls records/forms that must be maintained with the production unit, including work instructions, recipes, drawings, standard operating procedures, part programs, batch records, engineering change notices, and shift-to-shift communications, and edits "as planned" and "as built" information.
- **Data Collection/Acquisition**—Provides a link to obtain the intra-operational production and parametric data.
- **Labor Management**—Provides the status of personnel in an up-to-the-minute time frame.
- **Quality Management**—Provides real-time analysis of measurements collected from manufacturing to assure proper product quality control and to identify problems requiring attention.
- **Process Management**—Monitors production and either automatically corrects or provides decision support to operators for correcting and improving in-process activities.
- **Maintenance Management**—Tracks and directs the activities to maintain the equipment and tools to ensure their availability.
- **Product Tracking and Genealogy**—Provides visibility to where work is at all times and its disposition. Status information may include personnel working on it; component material by supplier; lot and serial number; current production conditions; and any alarms, rework, or other exceptions related to the product.
- **Performance Analysis**—Provides up-to-the-minute reporting of actual manufacturing operations results along with the comparison to past history and expected business result.

This book also provides a list of functions that would be included in a full MES implementation. Although this list includes the same system functions as indicated above by MESA International, it is based more on specific software divisions within a system and will be described using outlines from currently available software system products. These functionalities are divided into *core functions*, which are directly associated with managing production, and *support functions*, which include what might be called peripheral or support activities.

Core Functions

- **Planning System Interface**—This describes the connection with the planning layer.
- **Work Order Management**—This function manages work orders, including scheduling, for all orders in the system.
- **Workstation Management**—This function is responsible for implementing the direction of the work order plan, workstation scheduling, and the logical configuration of each workstation.
- **Inventory Tracking and Management**—The inventory tracking function develops, stores, and maintains the details of each lot or unit of inventory.
- **Material Movement Management**—The movement of material, manual or automated, is managed and scheduled through this function.
- **Data Collection**—This segment acts as the clearinghouse and translator for all information that is needed and/or generated on the plant floor.
- **Exception Management**—This function provides the ability to respond to unanticipated events that affect the production plan.

Support Functions

The functions identified in the following list are only a representation of possibilities; they do not constitute an exhaustive list of what is available or what will be on the market in the future. The idea is to provide the ability to "plug and play" support systems that currently exist in the facility or any new products that might be acquired in the future. The current most popular functions include:

- **Maintenance Management**
- **Time and Attendance**
- **Statistical Process Control**
- **Quality Assurance**
- **Process Data/Performance Analysis**
- **Document/Product Data Management**
- **Genealogy/Product Traceability**
- **Supplier Management**

Many plants have much of this functionality included in their daily activities and the production process. The MES, by definition, refers to the computerized, integrated, and on-line applications. Once again, the definition of a manufacturing execution system is **an on-line *integrated* computerized system that is the accumulation of the methods and tools used to accomplish production.**

References

MESA International, *The Benefits of MES: A Report from the Field*, MESA International, Pittsburgh, PA, 1994.

MESA International, *MES Functionalities and MRP to MES Data Flow Possibilities*, MESA International, Pittsburgh, PA, 1994.

2 | Computer Software Systems Used in Manufacturing Management

Business management has used computer systems for a number of years primarily for financial requirements. The orientation toward financial applications is one reason most systems relating to manufacturing are not as effective as might be expected. During the past forty years, manufacturing planning systems have evolved through Material Requirements Planning (MRP), Manufacturing Resources Planning (MRPII), and Enterprise Resources Planning (ERP) and have become an important part of company-wide management planning. These systems are frequently called manufacturing control systems or manufacturing planning systems. In this book, they will be referred to as planning systems that comprise the planning layer. There are also sophisticated computer and programmable logic control systems used to control devices on the plant floor. These systems comprise the control layer. Between the planning layer and the control layer is the execution layer, made up of systems and methods operating in the real world to accomplish production. In this chapter, these systems are briefly outlined and relationship between the planning layer, the execution layer, and the control layer is described.

- **Planning Layer**—Manufacturing resource planning
- **Execution Layer**—Manufacturing execution systems
- **Control Layer**—Device control logic

Planning Layer—Manufacturing Resource Planning

MRP systems were developed to provide better information to plan inventory quantities and costs. They are based on a master schedule that indicates the quantity of products by specific part number to be built in a given time period. Once the master schedule is available, this can be matched against the bill of material for each product to determine the quantity of material by part number that will be required. If the current inventory and lead time for each part number are included in the equation, a plan can be developed that shows when to begin production and how much capacity is required.

Manufacturing Resources Planning (MRPII) systems are a later development of MRP that broadened the application to include other company departments such as finance, marketing, engineering, purchasing, and information systems.

Enterprise Resources Planning (ERP) is the latest evolution and again broadens the system, usually including distribution, product data management, and supplier management. ERP systems are built around later information technology, including database management systems, client/server computer systems, and improved communication capabilities between systems such as CAD, product data libraries, and plant floor data collection devices.

Fundamental to all computerized manufacturing systems is the need for accurate part numbers and bills of material. There are other important parts required, but without part numbers and bills of material, there is no system.

- **Part Number**—Every item to be manufactured or that is a part of an item to be manufactured must have a computer-recognizable identification number that is unique to that item.
- **Bill of Material**—The bill of material lists each item used in the part to be manufactured by part number and quantity.

- **Routings or Process Steps**—Each item on the bill of material that is to be manufactured must have the manufacturing process outlined in routing steps, sometimes referred to as operations or operation codes. Operation codes are assigned to workstations and usually have time standard information included.

With this information, the computer begins the MRP run with the number of items to be manufactured for part number xxxx. The bill of material indicates that part number xxxx requires two of part number 1234 and four of part number 678. The bill of material shows that part number 1234 is a purchased part. The planning system indicates that the quantity of part 1234 in inventory and available for use exceeds the amount required for the current plan. Part number 678 is made up of part number 999 (raw material that is on hand) and requires two drilled holes. In our imaginary plant, all drilling operations are called operation code 222, and each carries a time standard of five minutes. The system indicates that to manufacture each part number xxxx, the computer calculates a material requirement of two 1234 (from inventory) and four 678 to be manufactured, requiring four units of 999 from inventory and eight units of operation code 222, or forty minutes of drilling time. This is a simple example of MRP, the basis for nearly all manufacturing planning systems.

An understanding of the details and the modules that make up these systems is helpful in seeing the relationship between the planning and execution layers. The modules included in this outline are generalized and are not completely described or representative of any system that any supplier might deliver.

- **Master Production Schedule (MPS)**—The master production schedule is an anticipated build schedule for end products to be manufactured. It tells manufacturing what and how many to make in a specific time period. It provides the basis for material requirements calculations, making customer delivery promises, and planning plant capacity utilization.
- **Material Requirements Planning (MRP)**—This module provides formal plans for each part number in the product bill

of material. Beginning with the product quantities in the MPS, the MRP module calculates the quantity for each item (part number) in the bill of material. After the quantity of each item required is determined and measured against inventory on hand, the quantity to make or purchase is calculated. The element of lead time is then added to the equation to determine when these items must be ordered or when to begin manufacturing.

- **Capacity Requirements Planning (CRP)**—CRP is a definition of the existing capacity to manufacture, usually expressed in some term of output. If a workstation can produce fifty units of a given operation per hour, then the capacity is that number times the number of hours utilized. Using that information as a base and adding the quantities required developed in the MRP calculation, the system can determine how much time for each operation is needed to meet the planned production schedule. This provides the queue length for each operation on an infinite (unlimited) loading basis. The assumption is that the capacity for the operation exists and is available at 100%. The objective in capacity planning is to measure capacity against planned production.

- **Shop Floor Control**—Shop floor control is used to measure where shop orders are in their routing. Although the information indicates the work has passed a specific point, there are a number of other determinations that can be made. Calculations include lists of work yet to be done for each order and operation code as well as lists of work that has been completed. If the information is current, this could be a good tool to manage work on the shop floor, but in most cases, the information is used to report what has been done. This information is a very useful tool to calculate inventory value and the execution of material plans; it can include detailed scheduling of individual jobs.

- **Purchasing**—Earlier, when discussing material requirements, there was mention of items in inventory and raw material. The acquisition of materials for manufacturing that fit the needs of the planning system is an obvious connection.

When the plan has been developed from the master schedule and the material requirements calculated, purchasing requirements are established.

■ **Inventory Control**—Within most companies, the amount of material in inventory is a major factor affecting company cost management. Inventory knowledge is equally important when planning manufacturing to ensure adequate material is available to meet planned production. Inventory quantities are usually maintained based on a beginning quantity plus quantities received minus quantities consumed for each part number that is in inventory. Quantities in inventory are usually indicated as an aggregate amount. An example is a part number that shows there are 50,000 available. Such data usually does not reflect details such as receipt date, supplier data, lot number, or specific location.

■ **Product Data**—All data supporting the products resides in this module. The data can include bills of material, routings, standards, process data, quality assurance standards, machine set-up times, part configuration, tool data, etc. This function may exist within the planning system or may be a separate software package that is accessed by the planning system.

■ **Cost Accounting**—This module provides for inclusion of cost and other data collection within the planning system.

Many manufacturing planning systems currently in place have only these or some of these capabilities. Newer systems might also include ties to time and attendance, accounts payable and receivable, order management, financial reporting, payroll processing, production scheduling and dispatching, etc. Newer systems might also measure planning horizons in hours, but most older systems have planning horizons of days or weeks, with little or no access to information or change until the computer department again runs the entire system. Many newer systems have the ability to make partial updates on some scheduled basis, but on-line or random access is generally not possible for the casual user. The general idea is that a plan has been made based on information current when the system was run, and manufacturing should take that information and do its job. Current data plus data collected between now and the next system cutoff date will be used for the next system run.

Most planning computer systems run in a batch mode, with only one function running and accessible at any time. Generally these systems do not allow manufacturing personnel to access the computer system directly. Information is most often provided in the form of printouts. From this information, plant management must determine its priorities and make products. Any changes or misinformation must be recognized by the line managers and responded to accordingly. Planning systems are usually reactive or reporting systems that can indicate a change has been made. They are not designed to be proactive to anticipate or react to daily plant floor changes.

The following description of linear programming for shoploading, from *Production Management, Systems and Synthesis*, might also apply when describing manufacturing planning systems:

> This is a technique of decision making under conditions of certainty. Those risk factors that exist, including machine breakdown, manmade interactions that produce variability and other disturbances are reserved for later interpretation and action by the production manager.

Control Layer—Device Control Systems

At the other end of the manufacturing process is the control systems layer; it makes process and machine functions occur. The control layer concerns inputs and outputs, or status points, of the process. These points can be relayed as they occur, trended as part of the functionality within the controls themselves, or stored in databases for analysis.

PLANNING SYSTEM

Master Schedule   Inventory   Material Requirements
Capacity   Purchasing
Product Data   Cost Accounting   Shop Floor

CONTROL SYSTEMS

Machine Tool   Oven Sensor   Storage/Retrieval System   Robots
Work Cells   Process System   Tool Storage   Conveyor System

Some examples of device control systems include:

- A system used to control the movements of one or more robotic devices
- An oven control system that includes temperature control, conveyor speed control, and exhaust air monitoring
- A quality assurance test station with the ability to retrieve specific product test requirements, monitor and collect test information, and store and/or distribute the results
- The molding machine controller used to control temperature, time, pressure, etc. based on information retrieved from a data library located at the machine or at a remote location
- A CNC (Computerized Numeric Controller) machine tool controller that converts a CAD program to the actual production of a part through motor control and measurement
- The process controllers that monitor and control products, flow rates, temperature, pressure, etc.
- An Automatic Storage/Retrieval System (AS/RS) controller that can manage inventory movement and maintain current inventory data for all products stored in the AS/RS system
- A SCADA (Supervisory Control and Data Acquisition) system

The systems used for equipment control can be very sophisticated, most often using programmable logic controllers (PLCs) or computers. Although some companies have the technical resources to build their own control systems, most often these systems are supplied by the process or device vendor with the necessary logic to accomplish the necessary functions. A few years ago, these systems were commonly referred to as "islands of automation." Many of these "islands" are highly sophisticated within their activities, but the big question was communicating between or with their control systems. There have been many products on the market that were meant to address this communication question. The most widely promoted was the Manufacturing Automation Protocol or MAP. MAP was meant to be and has become a common method of connection between computers and plant floor devices from different vendors. This, however, did not solve the major problem, which is not *how* to pass information but *what* to pass and *when*. The following are some *what* and *when* considerations:

- Engineering wants to locate all current work orders for a given product to determine the effect of an immediate engineering change order.
- Some purchased material that is specific to a given customer's order currently in process has arrived as a partial shipment, 72% complete. Where is the order and what schedule response is most appropriate?
- A process or machine critical to a product needs preventive maintenance. How are the current orders to be scheduled or rescheduled?
- A customer requires specific operator information including operator, date, and ambient conditions to be supplied for each item produced.
- There are 26 work orders totaling 443 hours of work for a specific routing location. What is the optimum sequence schedule for these work orders and what factors should be considered?
- The president of a high-volume customer has just called and needs to know tomorrow if he can double the quantity on the current order in-house without affecting the delivery schedule.

These questions or situations cannot be adequately responded to by the manufacturing planning system or by the control systems on the plant equipment. They are real questions that management encounters every day, and they require answers very quickly with the most accurate response possible. The answers lie in the execution system, the layer *between* the planning system and the device control system. The execution system communicates with the planning and control layers, translating data from both sources, interpreting that data, and passing the appropriate question or answer to the correct point of use. The easiest job of the MES is to process a question and deliver a specific answer. More sophisticated systems can process a series of questions and present or implement the answer automatically.

Figure 2.1 shows the manufacturing computer system hierarchy, including the planning system, the MES, and the device control system.

Figure 2.1

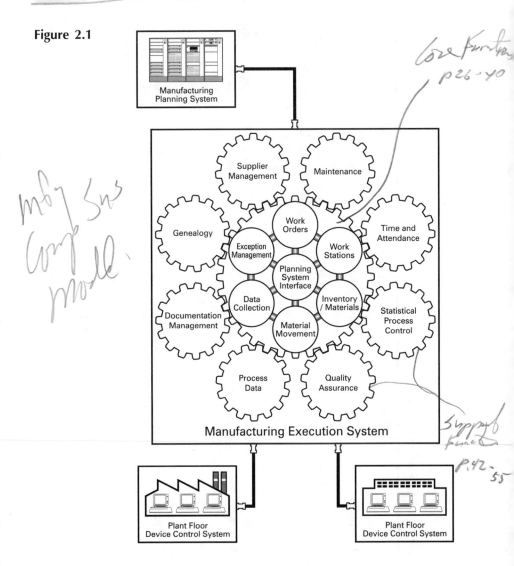

Core function p26-40

Support function p.42-55

System Application Examples

	CAD/CAM	Planning Level	Execution Level	Control Level
Inventory		Maintains aggregate. Triggers inventory data releases to increase inventory quantities.	Maintains detail inventory data and storage lists. Assigns inventory to specific orders. Tracks WIP through routing steps. Reports any changes or discrepancies to the planning system.	Retrieves inventory at the direction of the execution system.
Bill of Material	Prepares and maintains the current data.	Various uses in scheduling, requirements planning, inventory, purchasing, costing, etc.	Used to make inventory assignment and move decisions. Can be used to deliver material by routing steps.	
Routings	Prepares and maintains the current data.	The shop floor module uses routings to guide the dispatch list preparation. Capacity planning uses routings and time standards.	Does finite scheduling of production stations and directs the movement of material handling systems. Develops the work load by operation and workstation.	

Process Data	Prepares and maintains the current process information and instructions.	Retrieves and downloads information to the control system.	Sets the process control parameters. Examples include the pressure setting for an injection molding machine or temperature setting for an oven.
Scheduling	Provides gross scheduling and dispatch lists.	Schedules by production stations and material handling systems on a current event/current resource basis.	Performs the work according to the queue provided.
Capacity Planning	Infinite and total capacity planning.	Gives finite and actual current capacity by operation, workstation, work cell, and material handling system.	
Receiving	Matches inbound material to the purchase order.	Downloads QC test information and gathers QC data. Assigns inventory location and updates the planning system with all data requirements.	

System Application Examples (continued)

	CAD/CAM	Planning Level	Execution Level	Control Level
QC Testing	Establishes test criteria and reporting requirements.		Retrieves and downloads test requirements. Collects data and uploads to planning system.	Executes the instructions and provides test data.
Engineering Change Order	Generates change data.		Tracks ECO to match production orders.	

The idea for manufacturing execution systems is not new. As seen from this function outline, the coordinating effort between planning and the plant equipment has always been a part of manufacturing. The opportunity with an MES is to build a structure that provides consistent delivery of coordinated information that is current and available as necessary.

References

American Production and Inventory Control Society, *APICS Dictionary*, eighth edition, APICS, Falls Church, VA, 1995.

MESA International, *MES Functionalities and MRP to MES Data Flow Possibilities*, MESA International, Pittsburgh, PA, 1994.

MESA International, *The Controls Layer: Controls Definition and MES to Controls Data Flow Possibilities*, MESA International, Pittsburgh, PA, 1995.

Starr, M.K., *Production Management, Systems and Synthesis*, second edition, Prentice Hall, Englewood Cliffs, NJ, 1972.

Vollmann, T.E., W.L. Berry, and D.C. Whybark, *Manufacturing Planning and Control Systems*, Dow Jones-Irwin, Homewood, IL, 1984. This book is highly recommended as a source of information on manufacturing planning systems.

Wight, O., *MRP II: Unlocking America's Productivity Potential*, CBI Publishing, Boston, MA, 1981.

3 MES in Different Manufacturing Environments

In Chapter 1, MES was defined as an on-line *integrated* computerized system that is the accumulation of the methods and tools used to accomplish production. This has been restated to emphasize that all companies doing any kind of manufacturing or production have methods to accomplish the desired output. The point here is that those methods can most likely be made more effective when they are defined and integrated into a consistent model. This does not require a computer system, although a computer is generally the best method to assimilate and present the vast amount of data in a consistent and coherent manner. This presentation of data is likely to be an improvement over intuitive reactions that have to be made in a crisis environment, often referred to as "putting out fires." A well-managed production facility that meets production goals contrasts significantly with a crisis management approach that may or may not accomplish its goals.

One of the newest concepts in manufacturing is "agile" manufacturing. In a presentation by Kelly Thomas of Electronic Data Systems, agile manufacturing was defined as "delivering the right product, at the right time, the right cost, and the right quality with unplanned, unpredictable, continuously changing demand." He suggests this is a change from *economies of scale* to *economies of scope*. Scope in this context means reaching beyond the methods and needs of the supplier company into the needs of the individual customer. "Any color you want as long as it is black" no longer applies. In the real world, this means much broader product variety (note the variety and options in the automobile industry), immediate availability (mail-order personal computers), shorter production campaigns (batch processing), greater diversity of packaging and presentation formats (pharmaceutical industry), shorter product life cycles, better identification of customer needs, and the ability to fill those needs with products of a quality that is satisfactory to the customer *every* time a purchase is made.

Manufacturing execution systems can play a major role in creating a more agile company. Immediate information presented in the most meaningful way brings an improved focus to problems and better ways to exploit opportunities. Better information allows an improved picture of the variables involved in making immediate plant floor decisions, and through the computer some of these issues can be addressed proactively without taking valuable time from other management issues. MES can provide more data, more timely, more accurately, and more consistently to more people, providing a more sound base from which to react.

Other important ideas in manufacturing are continuous improvement and employee involvement through teams and other methods of empowerment. MES is a natural fit in any program where accessibility to current and accurate information is important to the participants. Where is that *not* appropriate?

The idea of MES is being applied to all manufacturing environments, from discrete-part manufacturing to process manufacturing such as a chemical plant, a pharmaceutical plant, or a brewery. Each application has its own MES variation, but the concept is the same. It would be difficult to list every type of business where MES is applicable, but the following industries are obvious candidates:

- Aircraft equipment manufacturers
- Appliance manufacturers
- Automotive and truck manufacturers
- Breweries and soft-drink processing
- Computer and communication equipment manufacturers

- Electronics equipment manufacturers
- Farm equipment manufacturers
- Food-processing companies
- Furniture manufacturers
- Large printing operations
- Pharmaceutical manufacturing
- Multiple-product chemical and petroleum processors

This list could include many more types of businesses. Most manufacturing/processing companies are required to make optimum use of resources and must consider many variables in the daily operation of the manufacturing facility. An integrated MES can be a substantial contributor.

Four manufacturing environments are suggested, each with substantial overlap:

- Discrete part
- Repetitive discrete part
- Batch process
- Continuous process

Within these definitions there are variations:

- **In-line**—In-line manufactured items are built following a defined, consistent route to workstations, such as an assembly line.
- **Defined routing**—The workstations are not necessarily in a line. Manufacturing follows process steps outlined in a routing defined by work requirements that necessitate the movement of material to varying locations.

In addition to variations, there are often hybrids of these classifications. For example, pharmaceutical production often contains elements of both batch and discrete environments in the same factory.

Almost by definition, the MES is a top-down-driven process, but the system can be adapted to fit specific needs that demand other than a top-down approach.

- **Plan and execute**—In this application, work is planned and directed to specific workstations with accompanying material and information. The operator, if there is one, has the work planned and the material delivered with few choices, if any, except to accomplish the next task.
- **Plan and have ready**—This application allows more freedom for the operator to choose which order to work on next; it requires the system to respond after the operator input is received.
- **Reporting**—Although MES is generally used as a proactive tool to cause improved results, in some cases the main function is to collect and report data relating to activities on the plant floor. This information is then displayed for the appropriate user or routed to the planning system. In FDA-regulated industries, one of the most important demands of MES is its ability to store records of production (device history records, batch records, etc.) with the appropriate access and security.

Although most applications will fall into one of these categories, it is not the intent to limit the types of systems or MES applications. This book is intended to define and explain MES in general terms, and throughout the text most application examples will focus on discrete-part manufacturing. Case studies of applications in various manufacturing environments are provided in Appendix A.

References

Goldman, Steven L., Roger N. Nagel, and Kenneth Preiss, *Agile Competitors and Virtual Organizations*, Van Nostrand Reinhold, New York, NY, 1995.

Maskell, Brian, *Software and the Agile Manufacturer*, Productivity Press, Portland, OR, 1994.

Thomas, Kelly, Electronic Data Systems, "MES in the Age of Agile Manufacturing," A presentation at MESA Roundtable 4, Chicago, IL, September 13, 1995.

4 MES Core Functions

The manufacturing planning layer and the control layer descriptions show there is very little relationship between them. The planning system works in hours, days, or weeks and is usually not on-line and available for change or inquiry. At the same time, the device control layer works in microseconds and is generally internal to and part of the functionality of the device, such as a machine tool. The MES fits between these two layers, on-line and able to communicate with the planning system and the device control systems.

The MES layer can be divided into functional parts. This allows pieces to be understood and developed in building blocks instead of one large massive software package. This also allows systems to be implemented and changed by segment as requirements change.

Although not every MES product is divided in exactly the same way, these functions are a part of nearly every production system in one form or another. They are considered as core functions because they are interrelated and basic to most production systems. Each of the functions outlined can be more extensive or, in some cases, omitted entirely, depending on the specific company, the amount of integration in place, and the manufacturing process. There is no suggested order of precedence in the functions. Your manufacturing practices will define your system. The core functions include:

- Planning System Interface
- Work Order Management

- Workstation Management
- Inventory Tracking and Management
- Material Movement Management
- Data Collection
- Exception Management

Planning System Interface

The MES should be directly coupled with the planning system to enable passing information between each system. Due to the large variety of installed systems, this interface is usually custom-developed software that fits the specific planning system and the MES. To define what information is to be transferred and what format and timing should be in place, collaboration with and between the MES vender and the planning system vendor is recommended.

This connection should provide the transfer of all data to and from the plant floor, usually starting with work orders. Other information can include inventory information, labor information, work order progress, etc. In the ideal system, planning system data exchanges and production information requirements should be provided by the MES. Designing the total manufacturing system with MES as the hub allows data to be stored and processed in the MES for use by any other user on the network. The general idea is to use the MES system as a client/server for all users and generators of data. When the data is in the MES data storehouse, the MES can assimilate and prepare information to meet all user requests that currently exist and those that are added later. Although fixed route communications from the place of generation to the place of use may seem a quick and easy answer to a specific request, invariably someone else will want that same information with a slightly different twist. Another reason for the centralization of information is future use. MES is a tool that will be evolving for many years, with better ideas for information use generated frequently. Information analysis and distribution over the long term can most easily be accomplished with the MES acting as the information hub and storehouse, allowing consistent development of information tools and access to the data storehouse.

The contribution of MES is not dependent on the existence of or a connection to a planning system. The plant floor that is functioning without a planning system will be improved with access to MES information since nearly all systems allow data to be manually entered into the MES. The tie between the two layers is similar to the tie between the planning layer and the machine control system. Each layer is or can be completely stand-alone, allowing the best development to fit the user objectives of each layer. The tie between the layers then enhances the operation of each.

Some of the issues regarding computer equipment and operating systems will be discussed later in this book. For now, it is sufficient to assume that the communication link can be made between the existing or proposed planning system equipment and the proposed execution system.

Work Order Management

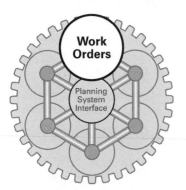

The MES accepts automatically or manually entered information that identifies what is to be produced and its quantity. This is usually done using work orders that designate the work order number, the part number of the item to be produced, the quantity, the requested completion date, and the method of prioritizing.

This module manages changes to orders, establishes and maintains schedules, and maintains a prioritized sequenced plan. It can also assign and unassign inventory to work orders. This may appear to be a redundant capability since the planning system may have provided a schedule when the work to be done was established. The role of the MES is to accept the information provided and manage accordingly until or unless it is necessary to respond to unplanned events. An example might be a machine breakdown that is critical to the highest priority order. While it might be difficult or impossible to query the planning system to determine action to be taken, the MES has on-line availability to respond, either by automatically rearranging the work order priority list, presenting production management with alternatives, or in the simplest version sounding an alert that will allow management to access current information and make the appropriate decisions.

A standard part of the MES is releasing orders to production and establishing a current order priority list based on sequencing rules or other schedule-producing methods such as simulation or constraints management techniques. In recent years, manufacturing management has seen significant improvement in scheduling work and tracking available resources, but there is much more to come. The current and future scheduling techniques are another important piece of MES and will be discussed in Chapter 10. The point to be made here is that they should be an integrated part of the MES, not a separate system.

Frequently, changes must be made to released orders. Within MES, order modifications and notations can be done easily, such as:

- Make schedule changes.
- Enter quantity changes.
- Make routing revisions.
- Split orders into smaller quantities.
- Combine orders into a larger lot size.
- Mark an order for material shortage.
- Place an order on hold or some other status.
- Append messages to orders.
- Run simulation analyses or "what if" queries.
- Establish or reestablish sequencing or prioritizing.

The work order management function maintains a constant real-time view of the work orders in the current backlog and the status of each order. Continuous-flow production systems would need different information that may require more or less functionality in this area.

Workstation Management

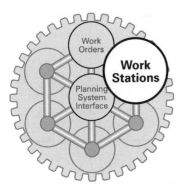

This function is responsible for implementing the direction of the work order plan and the logical configuration of the workstations; it can include the direct control interface and connection with each workstation.

The planning, scheduling, and loading of each operational workstation are done here. This module provides the current and total shop load by operation and/or workstation by using routing data and current time standards. Based on this plan, the system will request and manage delivery of inventory, tooling, and data in response to bill of material requirements. It will also issue and execute commands to move the required items to the planned workstations.

MES systems are highly configurable to each customer's operation and require building a logical model of the facility. This model begins with a listing of all departments, workstations, and operations in a facility. These are then matched to establish which workstations can do which operation or operations.

Workstation Operation Assignment

Department	Workstation	Operation	Description
ABC	0001	112	Hole punch
		114	Hole drilling
	0044	119	Welding
DEF	1230	220	Grinding
MNO	4422	460	Painting
XYZ	2210	999	Packing

Once this model has been established, the MES matches the routing for the required part number to develop the shop load by operation. If an operation is available at more than one workstation, rules within the model determine the loading accordingly.

Manufactured Part Routing
Part Number: 10044
Description: Welded angle bracket with two 1/4" holes

Step	Operation Code	Time	Set-up Time	Description
1	112	5	15	Drill holes
2	119	10	10	Weld brackets
3	1230	10	10	Grind weld
4	4422	5	30	Paint
5	2210	10	0	Pack

Functions within this segment of the system can:

- **Assign operation codes to work cells and workstations**—Each part number to be manufactured has a routing that lists each operation and the sequence of those operations. The logic in the MES follows the operation steps on the routing and matches those steps to actual workstations with the capability to perform those operations. Assignment of operations to workstations is part of the system management.
- **Optimize work orders**—For each workstation, a schedule is developed using sequencing rules, simulation techniques, or other planning methods currently in use in a facility. When the queue or current load for each workstation has been established, it may be necessary to determine in which sequence the work should be done. It is possible to have sequencing rules for each workstation and/or to do a quick simulation analysis to determine the optimum sequence for work in the current queue.
- **Establish the current plant load for each operation and workstation by using routing and time standard data**—By knowing

the part number, quantity, routings, and time standards for each order, the queue for each operation is established. This information can be presented on the computer screen with listings by order, due date, time required, or by nearly any other required display. In some plants, there may be a requirement to show the queue of those orders requiring a specific workstation as the next step and another listing that shows all orders with this operation in their routing. Once the logical model of the system has been created and the work to be done has been entered into the system, the plant load is established (for this moment), along with a sequence of the work. This serves as the basis for the work to begin. The appropriate material handling system (manual or mechanized) is then notified that specific material is required at a specific workstation.

- **Retrieve and download programs to plant floor devices** including robots, automatic insertion machines, paint booths, automatic valves, computer numerical controlled machines, and quality assurance test stations. Again following the schedule that indicates the sequence of work for each workstation, the MES can automatically retrieve information and direct that information to the specific workstation to match the sequence of work orders.
- **Maintain a current map of workstation and operation availability**—If an operation becomes unavailable, the MES should respond automatically to reschedule work into other available resources or sound an alarm for appropriate management action. Depending on the degree of integration in the MES, unavailability might include machine breakdown, scheduled machine maintenance, unavailable tooling or labor skills, etc.

This module maintains a picture of the workstation resources that are currently available and their backlog of assigned work. It also causes events such as material movement or schedule changes to occur according to the work order management plan.

Inventory Tracking and Management

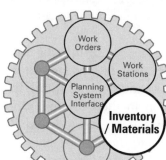

The inventory tracking function develops, stores, and maintains the details of each lot or unit of inventory, including the current location. For purposes of the MES, inventory includes anything that is needed for production: tooling, fixtures, raw material, work-in-process, drawings, specific labor skills, and any other item that could be listed on the bill of material. For a number of reasons, it would be very difficult for most planning systems to carry the detail and be able to respond on-line to issues regarding inventory. Information that resides in the planning system (usually aggregate data) is requested by the MES, and changes in inventory that need to be known at the planning level are supplied by the MES when they happen or on an established updating schedule. Although many planning systems have extensive inventory information capabilities, the local inventory data must either be on the MES or immediately available on-line when events cause any change to plant floor priorities.

This segment functions in the following specific ways:

- **Manages, directs, and controls all raw material and work-in-process inventory** to be able to provide a specific item of inventory to a workstation in response to the order sequence schedule. The actual management of the material is done here, acting on information stored in the MES, in the planning system, or in an inventory storage device such as an Automatic Storage/Retrieval System (AS/RS). The MES should have stored in its database or available on-line whatever information is necessary to choose a specific inventory item. This might include current location (in-house or at a vendor), date received, lot number, quality information, and any other information that forms the decision of which to choose. In-process inventory information is also maintained or available at the MES level.

■ **Locates and retrieves all supporting material and information** in response to the production plan. In this context, inventory includes anything that is required to accomplish production of a specific order. This includes drawings or other technical information (CAD data), tooling and fixtures, specific labor skills, etc.

■ **Maintains and provides access to detailed information for each item**, such as ID number, location, quantity, date received, vendor, purchase order number, etc.
  □ Inventory listing by part number and location
  □ Inventory listing by part number, purchase order, and receipt date
  □ Inventory listing by part number, location, and work order assigned

One example of inventory management in the MES might begin with receiving material and processing that material to stock. When material is unloaded, a bar code label is scanned or a purchase order number is entered into the system. The system retrieves and displays a copy of the purchase order along with processing information. The processing instructions are completed and the material is moved to a specific storage location or to the next processing location. As the material is processed toward acceptance and available use, the MES captures data from each step, determines and displays the next operation, and updates the planning system accordingly. There are countless possible examples of using MES for improved inventory management. Let your imagination run wild. Inventory management is a significant area of MES contribution, and nearly anything you might want to do can be accomplished.

Material Movement Management

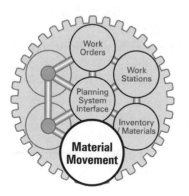

The MES has so far established the work order sequence, the workstation to do the work, and the inventory to be used. The system is now ready to cause the movement of specific inventory to the workstation. The system logic is based on the idea that everything has an identified location and must be moved from that location to the desired location. This really is as simple as it sounds when *accurate information* exists. Material movement is a non-value-adding function; in its simplest form, it means "take this item from this location and put it in that location." With the model of the physical facility built into the MES, the system must simply determine when to move what from where to where and issue the appropriate instruction. These instructions can be for manual movement or to sophisticated control systems that manage guided vehicle systems, automated storage/retrieval systems, or piping systems in process plants.

- Issue move tickets to a fork truck driver.
- Issue release information to an outside vendor.
- Advise the AS/RS control system to deliver a specific pallet of material from a given location or simply request the material, allowing the AS/RS control system to determine the appropriate retrieval location.
- Advise the conveyor system programmable logic controller to deliver material to a workstation.
- Advise the automated guided vehicle system to pick up at one workstation and deliver to another.
- Open a valve and turn on a pump to deliver liquid material to a specific location.

Data Collection

The data collection function is the "eyes and ears" of the MES, allowing the system to remain current. The application can be as simple as bar code scanning or as sophisticated as a data storehouse that collects and compiles large quantities of data. This segment of the MES acts as a clearinghouse and translator for **all** information that is needed and/or generated within the production facility. (This may not always be possible, but it is a very worthy objective.) This can include direct input of data being generated at upstream supplier locations within and/or outside a company.

Through various kinds of sensing devices and control interfaces, data from the shop floor operations can be collected, collated, and dispersed on whatever basis is desired. This is the primary method for all personnel to communicate with the MES. Plant floor data such as labor variance, statistical process control, time and attendance, lot traceability, and genealogy reporting is processed here. Another way to view data collection is as a vast data storehouse that maintains collected information for later compilation as desired. Some data sources include:

- Bar code scanners
- Voice-encoded information
- Radio frequency transmitters
- Programmable logic controllers
- Time and attendance systems
- Quality assurance systems
- Machine and process monitoring
- Other computer systems
- Manual data entry

The central premise is that data collection within the MES is the communications hub that is equipped and assigned to collect and supply the electronic and manual data between the plant floor and all users.

Exception Management

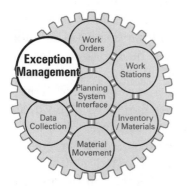

The most custom function of an MES is its ability to respond to unanticipated events that affect the production plan. What happens when a machine breaks down? How do you respond when material does not arrive on time? How do you react to customer order changes? Many of the responses to these unpleasant, unplanned, and everyday events typical in most manufacturing plants can be handled automatically by the MES. The MES should be able to take many changes in stride and implement alternative actions.

- The system can reschedule production or use alternate routings in response to a manufacturing resource loss.
- The system can monitor quality assurance information and adjust machines and processes to conform to specifications.
- At a very minimum, the system must inform management of exceptions and allow the response to be entered in the system.

Some companies might use the term "expert systems" to refer to exception responses. This suggests that some complex formula and analysis must be developed. The complexity is no greater than whatever a company's current methods might be as long as they can be defined and converted to action and logic steps to be implemented by the MES. For example, if company procedure is to place a hold on orders that are short material and exclude these orders from today's consideration, this can easily be done within and by the MES. Many other exceptions can be just as easily managed.

Scheduling

Scheduling is a critical component of manufacturing; therefore, it is an important part of MES. The real question is, "What is scheduling?" Scheduling can be forward or backward, finite or infinite, as simple as assigning a priority number or as complex as a computer simulation. Work sequencing rules are part of scheduling, as is the newer concept of constraint scheduling. In Chapter 10, both simulation modeling and constraint management are discussed in more detail. The important point here is that scheduling, no matter how sophisticated or simple, is a part of the MES and should be integrated into the system. In some instances, scheduling may reside within the MES package or be on-line as a separate node on the network that is accessed, as necessary, to provide schedule information to the core functions of the MES. Examples include scheduling work orders, scheduling into individual workstations, scheduling or assigning material to orders, scheduling the movement of material, and using on-line scheduling revisions as part of exception management that responds to unplanned resource interruptions.

Data Library

Within every manufacturing system there must be an information source or data library that provides information to support production. This information may reside in the MES layer, at the planning layer, in various locations (local or distant), or within sophisticated Product Data Management (PDM) computer systems (discussed in Chapter 5). In some systems, the information might be downloaded from the planning system along with other work order information. In other instances, the MES might automatically query the data library for information using the part number of the item to be manufactured. Typical information available in the data library includes routings, time standards, process data, tool data, quality control test information, set-up times, product recipes, and bills of material. The issues of where the data resides and how the MES has access are normal parts of the system requirements design process.

A System of Integrated Functions

The following system is an example of how MES can be effectively applied to reduce work-in-process and improve customer response time. A hypothetical company makes replacement printed circuit boards for military electronic equipment. It has 400 different finished part identification numbers. Each part number has a bill of material with part identification numbers and a routing with the required operations. The boards are manufactured to order with a maximum lot size of ten and an average lot size of three. There are usually 1,000 orders in process, each with an average of six routing steps. The assembly area has forty-eight workstations working two shifts. A piecework incentive system is in place, with varying values for each product. The company has a contractual requirement to provide genealogy information for each board and wants to improve the delivery time performance, both to shorten the delivery cycle and to meet delivery promises.

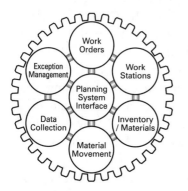

The production department receives work orders from the planning system weekly. Kitted material is received in tote pans that are grouped by the work order number. The tote pans are bar coded to provide tracking information of the tote pan/work order and are moved by an automated conveying system. All work-in-process is stored in a storage/retrieval system to be dispatched to an appropriate workstation automatically by the MES. Each worker clocks in through the MES bar code data collection system at each workstation.

The operation begins with a download from the planning system. The information provided includes a work order number, the quantity to be made, the part number to be made, the delivery week, and the priority of this order based on a scale of one to nine. The MES has a data library that includes routings and time standards for each part number. When the kitted material is received from the material warehouse, the tote pan ID number is matched to the work order number, put into the material handling system, and scheduled into the first operation on the routing. Depending on the sequencing rules for that workstation, the material is automatically dispatched and conveyed to an available workstation with the capability to perform the operation. The operator scans the bar code on the tote to indicate work is beginning on this order. The MES records the start time and the operator number. When the operator is finished, the work order number is scanned and the tote reentered into the material handling system to be returned to in-process storage or sent to the next operation.

The system maintains a model of what has been done, what is being done, and what is to be done. It uses current information about the location of each item in the production facility, and it determines movements to be made into and out of each workstation.

Some of the data that can be displayed by the system include the following:

- Each work order in the system by location
- Each work order in the system by part number
- Each work order by priority
- Each work order by due date
- Each work order with a material shortage
- Each work order by status
- The backlog at each operation
- The operations assigned to each workstation
- The current work order at each workstation
- The routing of each work order and a method to change the routing

- The variance from standard by operation and operator for each work order
- The current storage location by part number
- The personnel signed in for each location
- Time and attendance information
- Personnel working on each item on each order and when

Many functions are performed automatically or by an operator:

- Issue manual material move instructions
- Automatically receive totes from a workstation and place in work-in-process storage
- Automatically retrieve totes from work-in-process storage and deliver to a workstation based on the order priority number assigned and due date
- Automatically move totes between workstations according to the routing
- Supply labor reports
- Split and combine work orders by the system operator
- Maintain a message log for each work order
- Automatically update the planning system during the third shift
- Automatically print bar code labels to match new order entry or by operator request

Some systems will require all of the core functions, others will need only some, and in some instances it may be necessary to begin small and add functions later. Keep in mind that your system software divisions probably will not match these functions exactly.

5 MES Support Functions

T he previous chapter discussed the core functions that deal primarily with the actual management of the work orders and the manufacturing resources. This chapter describes other functions that should be part of the manufacturing execution system. They are no less important, but they seem one dimension removed from production.

There can be any number of support functions, and new ones are sure to be developed in the future. Those identified here may be more than your system will ever need, and in some cases support functions not included in this list may already be in place. Those presented are the most common and are shown to give an indication of the possible breadth of the MES application. MES is a continuously evolving part of manufacturing systems, and even if you do not need some function today, it is wise to keep your options open and consider where you might be in five, ten, or fifteen years, since it is very likely there will be new and better ideas for system applications in the future.

The support functions included are:

- Maintenance Management
- Time and Attendance
- Statistical Process Control
- Quality Assurance/ISO 9000

- Process Data/Performance Analysis
- Documentation/Product Data Management
- Genealogy/Product Traceability
- Supplier Management

Maintenance Management Function

Maintenance management is a big job in any plant, and being able to stay on top of resource availability and planned downtime can be a significant aid to production. Although this may seem obvious, schedules are frequently developed without an awareness of maintenance issues.

Maintenance management systems can be very sophisticated integrated packages that provide historical, current, and planned maintenance events. One system is divided into eight modules to track and schedule maintenance tasks:

- **Equipment Module**—This module maintains a complete history for each piece of equipment, including technical data, spare parts information, and running records of all maintenance performed on each equipment item.
- **Work Order Module**—In this module, all preventative and unscheduled tasks are assigned and tracked.
- **Preventative Maintenance Tasks Module**—This module generates and tracks every scheduled task, for every piece of equipment.

- **Statistical Predictive Maintenance Module**—Using statistical process control methods, the system tracks dynamic equipment data such as vibration, temperature, electrical load, or clearances to help predict equipment failure and prevent downtime.
- **Inventory Module**—The system tracks spare part inventory usage, stock quantities, and physical location.
- **Purchasing Module**—This module is the connection to vendors, receiving quotes, approving purchases, and managing the purchasing and receiving functions for maintenance items.
- **Labor Module**—The system tracks human resources by individual and by craft.
- **Analysis Module**—In this module, reports and analysis of maintenance data are generated.

Time and Attendance Function

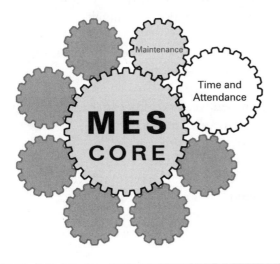

The collection of data regarding attendance is common in nearly all manufacturing plants. This information can be as simple as data from a time clock or as sophisticated as a badge scanning system. Once the data has been created, the MES can use the information for immediate production reasons and/or send it to other system users such as payroll and costing. There are many standard software systems on the market that produce time and attendance information which can be integrated and used on a wider scale. Since some personnel information may be sensitive, systems usually include built-in security locks to prevent unauthorized use of specific information.

The following product features are typical of time and attendance systems:

- **Time and Attendance**
  - ☐ Maintains employee master file with employee name, employee number, plant, department, default schedule, work center/group, direct/indirect, permanent/temporary, reporting assignment, accrued vacation time, and sick time.
  - ☐ Allows employees to select their own personal identification number (PIN) for their electronic signature.
  - ☐ Assigns employees to unique schedule and reporting groups. Each group has its own set of flexible, table-driven parameters for compliance with union contracts, work rules, policies, and schedules.
  - ☐ Provides for fixed, rotating, and flexible schedules.
  - ☐ Captures clockings on shop floor terminals, full screen terminals, and telephones.
  - ☐ Provides automatic log-in, log-out, and break features to reduce the number of transactions.
- **Labor Collection**
  - ☐ Supports discrete, repetitive, and process manufacturing.
  - ☐ Maintains a database of active work orders.
  - ☐ Validates transactions at the time of entry.
  - ☐ Supports single, batch, and multiple operations in process.
  - ☐ Automatically posts labor for indirect employees.
- **Supervisor Administration**
  - ☐ Has full screen access to one or more department groups via unique password for employee review, override, and reporting.
  - ☐ Performs employee review on a daily and/or weekly basis.
  - ☐ Generates standard supervisor reports.

■ **Payroll/System Administration**
  ☐ Provides permanent and temporary employee badge maintenance.
  ☐ Has interfaces to most payroll, human resource, and planning-level packages.

Statistical Process Control Function

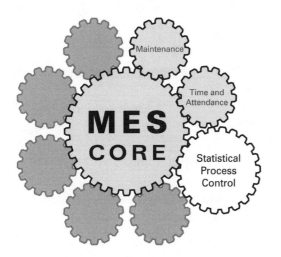

Statistical process control (SPC) is a quality control method that focuses on continuous monitoring of a process rather than the inspection of finished products, with the intent to achieve control of the process and eliminate defective products. It is a collection of tools, mostly statistical, which help to understand what is going on in any process. The tools usually included in SPC information are:

■ Flowcharts
■ Run charts
■ Pareto charts and analysis
■ Cause-and-effect diagrams
■ Frequency histograms
■ Control charts
■ Process capability studies
■ Acceptance sampling plans
■ Scatter diagrams

Quality Assurance Function

Quality assurance applications may or may not be tied together with SPC capabilities and/or ISO 9000. Either as separate packages or combined, they can easily be a part of the MES. Typical features included in quality assurance are:

■ **Receiving Inspection**
  ☐ On-line instructions for inspection
  ☐ Automatic statistical analysis
  ☐ On-line access to results by other functions
■ **Non-Conformances**
  ☐ Customer-defined types with varying methods of operation
  ☐ Trend and Pareto analysis reports
  ☐ On-line processing and assignment
■ **Supplier Rating**
  ☐ Generation of monthly summary letters
  ☐ Automatic data collection from other modules
  ☐ Detailed reports consolidating receiving, non-conformance, and corrective action information
■ **Corrective Action**
  ☐ On-line generation and processing
  ☐ Internal and external support
  ☐ Follow-up and late notices
■ **Gage Tracking**
  ☐ Tracking of both serialized items and consumable material

- ☐ Checkout and return
- ☐ Usage reports by work order, user, and serial number
- **Calibration Control**
  - ☐ On-line entry of results
  - ☐ Drift and trend analysis
- **Statistical Process Control**
  - ☐ Charting to character-based terminals or color graphic workstations
  - ☐ Rapid criteria set-up
- **In-Process Inspection/Test**
  - ☐ On-line generation of inspection criteria
  - ☐ On-line access to work-in-process status
- **Serialized Test/Inspection**
  - ☐ Rapid collection of test and inspection results
  - ☐ Complete life cycle result for each serial number

Process Data/Performance Analysis Function

Process data collection and management can be a standard package developed for specific applications such as time/cost variance information or process information such as SCADA. These systems may be purchased as a package or developed in-house with information collected through the data collection function. In any case, the information is a part of the MES and should be integrated and available throughout the system.

Documentation/Product Data Management Function

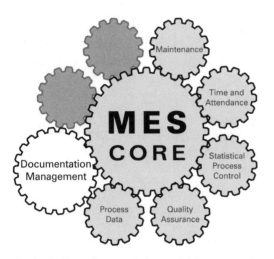

Product data on items to be manufactured has been available on computer systems since the early implementation of planning systems during the 1950s, when basic information such as routings and time standards was stored in a data library and used to make MRP calculations. Other information in the data library might include bills of material, tool data, quality assurance test data, machine set-up times, part configuration data, and CAD/CAM drawings and programs. Managing this information as well as receiving and archiving plant-floor-generated information is included in this portion of the system.

The amount of information necessary to manage manufacturing grows every day, and the idea of written information and manual data entry or manual delivery of information (to or from the plant floor) is too limiting. Modern product data or document management systems are excellent tools that provide immediate and accurate information wherever needed. Think of all the information (drawings, time standards, routings, process recipes, ISO standards, regulatory requirements, assembly instructions, test instructions and data, quality assurance parameters, electronic mail, etc.) that exists in your company and how it might be used by manufacturing at all levels, including the workstation level. The following are some examples:

- Transmit drawings or electronic data to workstations on request or automatically in-line with the production schedule.

- Send full-motion video and voice data to workstations or create this information at the workstation and append to the work order or production unit.
- Create production messages for reports and filing.
- Download process recipes to equipment on request or automatically.
- Activate engineering change orders on an immediate global basis.
- Automate document updating throughout the company.
- Validate process information for applications such as ISO 9000 or regulation compliance.

Communicating this information can be as simple as using a video screen that uses a GUI (graphical user interface) or a man–machine interface or automatically loading a CAD/CAM program into a machine tool programmable controller. The ability to communicate between the individual worker and the library of available information is becoming more economical and available through standard systems that are easy to install and easy to operate.

Product Data Management (PDM) is a general extension of techniques commonly known as engineering data management, document management, product information management, technical data management, technical information management, image management, and other names. PDM provides a common term, encompassing all systems that are used to manage product definition information. PDM systems and methods provide a structure in which all types of information used to define, manufacture, and support products are stored, managed, and controlled. PDM is used to work with electronic documents, digital files, and database records. These may include:

- Product configurations
- Part definitions and other design data
- Specifications
- CAD drawings
- Geometric models
- Images (scanned drawings, photographs, etc.)
- Engineering analysis models and results

- Manufacturing process plans and routings
- Numerical controlled part programs
- Software components of products
- Electronically stored documents, notes, and correspondence
- Audio and live video annotations
- Hard copy (paper-based and microform) documents
- Project plans

In short, any information needed throughout the life of a product can be managed by a PDM system, making data accessible to all people and systems that have a need to use them.

Typical elements of a PDM system include:

- **Data Vault and Document Management**—Data vault and document management provides secure data storage and retrieval of product definition information. The vault contains either the data itself or information that points to the actual location of the data.

- **Workflow and Process Management**—Workflow and process management can define and control changes to product configuration, part definitions, other product data, data relationships, and data versions and variations. Workflow and process management defines and controls the process of reviewing and approving changes to product data. The workflow and process are defined in terms of a sequence of events that must occur before modified product data can be released.

- **Product Structure Management**—Product structure management facilitates the creation and management of product configurations and bills of material. As configurations change over time, the PDM system tracks versions and design variations. Typical product structures contain attribute, instance, and location information in addition to standard bill of material data. These data enhance the value of the structure for activities outside of manufacturing planning. Standard bills of material can be generated automatically from the product structure.

- **Classification**—Classification of parts allows similar or standard parts, processes, and other design information to be grouped by common attributes and retrieved for use in products. This leads to greater product standardization, reduced redesign, savings in purchasing and fabrication, and reduced inventories.

- **Program Management**—Program management provides work breakdown structures and allows resource scheduling and project tracking. Resources and managed data are linked to provide an added level of planning and tracking. A key advantage stems from the ability to relate the work breakdown structure tasks to the PDM system's knowledge of approval cycles and product configurations.

- **Communication and Notification**—On-line, automated notification of critical events means that all personnel are informed concerning the current state of the project. Electronic mail is used to notify people about important events or required actions on-line.

- **Data Transport**—All data is stored and accessed under control of the PDM system; thus, a user need not know where in the computer network data is stored. The system keeps track of data locations and allows users to access data by a data set name. Moving data from one location to another is an operation the PDM system performs. Users do not need to be concerned with operating systems and network commands.

- **Data Translation**—The system administrator can predefine translators to be used to convert data between pairs of applications and to formats for various display and output devices. Triggers can cause data to be translated automatically from one application to another at appropriate times. Thus, the correct information is more likely to be used in any situation.

- **Image Services**—Raster, vector, and video images are treated the same as any other data by the PDM system. On-line access is provided to a wide range of previously difficult to distribute product information, providing this information in a structured manner to more users.

■ **System Administration**—The administrator sets up the operational parameters of the PDM system and monitors its performance. Administrative functions include:

  □ Access and change permissions
  □ Authorizations
  □ Approval procedures
  □ Data backup and security
  □ Data archive

Systems can be tailored to conform to corporate standards and to improve the efficiency of operations for individual users. In addition to user interface customization, the operational features of PDM systems can be tailored. For instance, the approval process may be set to require sequential approval by several individuals in one case and to allow a majority vote of all approvers in another. Systems can be tailored in many ways, including:

■ User interface layout
■ Modifying system messages and terminology
■ Integrating third-party applications
■ Adding new functionality

The connection and integration of PDM into the MES can have a dramatic and positive impact on production plan execution by providing a wide variety of information to the right place at the right time in whatever format best fits the user.

Genealogy/Product Traceability Function

Genealogy information is becoming an increasing necessity for many manufacturers with a need to backtrack through the manufacturing process for information on components or conditions for a serialized product or lot. The information can be detailed to whatever level is desired to know either the serialized units that contain a certain part or operation or the specific parts or operations of a

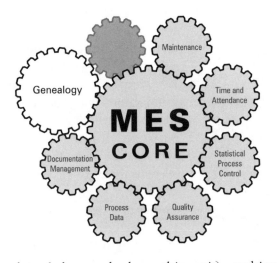

serialized unit. The most frequent use of this information is for warranty and product statistical information, as well as inventory use information.

For example, suppose a consumer products manufacturing company plans to collect information from over 300 data points in its process: personnel data (who worked on this unit), ambient and equipment conditions, and assembly material supplier information. The data, archived by the manufacturer on a chip installed in the finished product, can be accessed by a service technician to use for diagnostics and repair. This program provides the quality assurance department with a vast statistical base for analysis and gives purchasing much better supplier product performance information. Less sophisticated examples include backflushing inventory adjustments and labor variance reports by serialized unit or lot.

This is an area of growing significance in most manufacturing companies, and many MES applications have some basic built-in traceability functions. A requirements definition should be specific about which data to collect and how the information is to be collected—automatically through data collection equipment or manual entry.

Supplier Management Function

As outsourcing and just-in-time inventory management become more common, so too is the need for current information to stay abreast of upstream operations within a company or suppliers' operations. The degree of synchronization within all of the off-site sources is

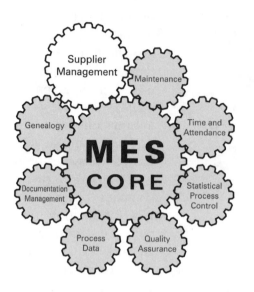

no less important than knowing the status of work-in-process in departments within a plant. This data can be supplied via on-line connections with suppliers' MES systems and/or through input devices within their manufacturing systems that provide on-line data. This degree of communication tie-in is new but is expected to grow substantially during the next few years.

MES is the on-line *integrated* computerized system that is the accumulation of the methods and tools to accomplish production. The key word here is

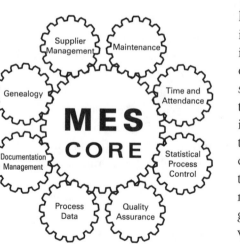

integrated. Most companies have systems as described in this chapter and are using these tools today. The on-line integration of these systems or, at a minimum, the accessibility of all of the information available in these systems will significantly enhance the ability to respond to and control manufacturing issues. Integration as suggested here will vary from one user to another depending on how accessible each existing software system might be. For example, it will definitely be more difficult to tie a ten-year-old maintenance system into a current scheduling system. Each instance must be examined to determine how and what can be done most effectively. At a minimum, the on-line user of the system should have one terminal that provides access to all system modules and appears seamless. As the system is developed, the modules should automatically interact, initiating and providing direction based on internal and external stimuli.

References

Maintenance—Datastream Systems, Inc., 50 Datastream Place, Greenville, SC 29607.

Time and Attendance—Applied Automation Techniques, Inc., 5753 Miami Lakes Drive, Miami Lakes, FL 33014.

Quality Assurance—Applied Automation Techniques, Inc., 5753 Miami Lakes Drive, Miami Lakes, FL 33014.

Documentation—Documentum, Inc., 5671 Gibralter Drive, Pleasanton, CA 94588.

Genealogy/Product Traceability—Industrial Computer Company, 5871 Glenridge Drive, Atlanta, GA 30328.

Product Data Management—CIMdata, 3909 Research Park Drive, Ann Arbor, MI 48108.

6 System Configuration/ Architecture

The system architecture (computer equipment relationship) is likely to be different for each manufacturing execution system application. Systems can range from one personal computer with a display screen to a hierarchy of computers on a worldwide network of nearly any configuration. Applications can be one department of a plant, one plant with many departments, or departments within plants within companies. There is no limit to the possibilities—whatever the needs, it is likely they can be met. When designing a system, keep in mind the purpose or purposes behind the design. Is the intent to provide information for management decision support or is the focus on plant floor usage to affect a given shop order or production step? There is a difference, and the question must be examined at the requirements development stage.

Most computer applications fall into two categories: decision support systems (DSS) or on-line transaction processing (OLTP). OLTP systems are used in many areas of business, particularly where an immediate response time is necessary. The reliability and availability of the system must be very high, with information that is consistent and correct. DSS systems are used to analyze data and create reports. These systems are generally not time critical and can tolerate slower response times.

Manufacturing execution systems definitely fall into the OLTP category and must provide up-to-the-minute on-line information in real time. The following table, taken from *Essential Client/Server Survival Guide*, provides some ideas of system differences:

Feature	OLTP Database Needs	Decision Support Database Needs
Timeliness of data	Needs current value of data.	Needs stable snapshot of data frozen in time.
Frequency of data access	Continuous throughout workday. Work related peaks may occur.	Supports knowledge workers and managers.
Data format	Raw captured data.	Multiple levels of conversions, filtering, summarization, condensations and extraction.
Data access pattern	Multiple users continuously updating the production database.	Mostly single user access.
Can data be updated	Current data is continuously updated.	Read only, unless you own the replica.
Performance	Fast response is a requirement. Highly automated, repetitive tasks.	Relatively slow.

Manufacturing execution systems come in any shape your system requirements might define. In most cases, the configuration will not be static since most MES systems are not an event but a continuously evolving process. Within this process, there will be application changes and system equipment changes.

To provide an illustration of possible system configurations, let's begin with an assembly cell with ten workstations with work order data entered into the system manually through the terminal by the area production manager (Figure 6.1).

Figure 6.1

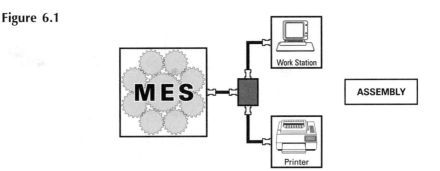

Add to the base system a planning level interface with download/upload capability, a robot, and a process system (Figure 6.2).

Figure 6.2

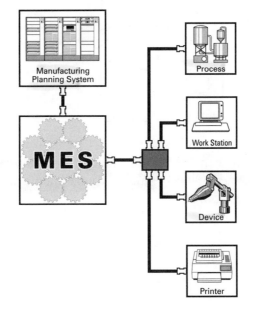

Enlarge the system to include off-premise CAD/CAM or a Product Data Management system, inventory management (mini-load AS/RS), and an off-site just-in-time material supplier (Figure 6.3).

Figure 6.3

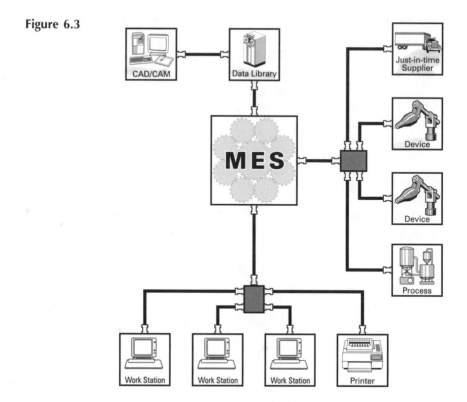

The system can be further enlarged to include multiple plants within the system having multiple departments with any variety of workstations (Figure 6.4).

Figure 6.4

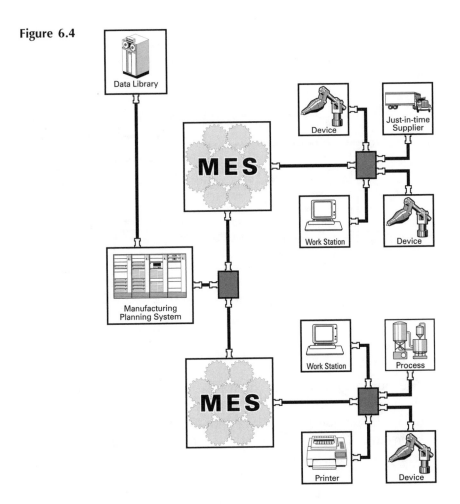

Figure 6.5 shows a hierarchical arrangement of MES applications. In this example, one MES can supply information to higher or lower level systems and/or communicate directly with the planning level.

Figure 6.5

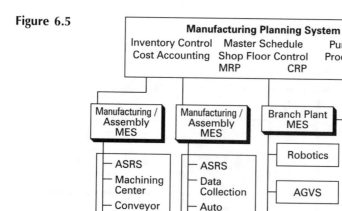

The purpose of these examples is to illustrate that there is no single answer or any limit to the possible application or arrangement of MES systems, and that is how it should be. The only real limit on an MES might be from the return-on-investment point of view and your imagination.

Reference

Orfali, Robert, Dan Harkey, and Jeri Edwards, *Essential Client/Server Survival Guide,* Van Nostrand Reinhold, New York, NY, 1994.

7 Equipment Used in MES

A fter developing the system requirements definition, equipment considerations will have to be addressed. Equipment issues in a system can include a computer or computers, communication equipment including telephone lines and local area networks, data entry methods and equipment, and product identification methods. This book provides an overview of many of the considerations and options in MES applications; the details of equipment application and selection should be provided by the vendors you are considering.

Computer Hardware

Equipment Platforms

Most computer equipment manufacturers have taken a very strong interest in manufacturing execution systems and have a variety of products to address the market. Selecting a vendor or vendors should center on those companies best able to serve the MES application over a number of years. This also applies to the more technical products such as operating systems and database management systems. The day is coming when it will be possible to connect most equipment and software products together randomly and they will operate correctly as expected. This is usually referred to as "plug and play" and can be seen in newer personal computers and operating systems, such as Microsoft's Windows 95, which can accept many hardware and software packages without additional internal programming. This will make systems procurement significantly easier, but until then, it is necessary to analyze how the components will fit together to meet the requirements definition.

It is beyond the scope of this book to make recommendations or address system questions such as operating systems, database management products, or specific equipment such as mainframe, workstation, or personal computer. Each of these, along with a number of other subjects included in information systems, depends largely on company-wide system applications and standards. The one thing that appears certain is that greater computing power will continue to be delivered in smaller machines and at lower cost.

Equipment Sizing

The system design will provide direction in selecting the size of the computer (mainframe, midsize, workstation, or personal computer) to be used in the system. There are at least three basic elements to system size requirements:

- Data storage capacity
- The number of peripheral devices such as terminals, shop floor devices and systems, and printers
- Computational requirements (quantity and speed)

The data storage capacity depends on the number of system records, the database management system, software programs to run the system, and software to run the computer equipment. The second aspect of system sizing is based on the number of interactive terminals to be supported and the number of non-terminal devices that require a response. The third is dependent on the breadth of the application: the number and rate of transactions that will be required of the system.

In addition to these considerations, there is the assumption of correct system design, which has a major effect on sizing requirements. The correct design includes the software program design, the database design, the correct computer architecture, and the correct functional assignments to equipment in the system.

Correct system equipment sizing and design should be the responsibility of the vendor. The buyer is responsible for defining what the system is expected to do; it is up to the vendor to supply a system that meets those objectives.

Reliability

Whenever an on-line system is planned, it is assumed the computing process must be continuous and backward supportive (previous events are required to support future computer functions). It also seems quite obvious that a work force standing idle while the "computer is down" is not acceptable. Some form of backup should be part of the system.

One method is to use a "fault-tolerant" computer system. This is the most expensive and reliable method and is popular in on-line transaction processing (OLTP) systems where service interruption could be very costly. The process is accomplished through duplicate hardware components, two systems running as one. If one system incurs a fault, the other will diagnose it, initiate a fix, and carry on the normal functions without interruption. Most computer equipment suppliers offer methods to address the problem of failure in other ways. For example, IBM offers a "high-availability" strategy that includes increasing levels of availability starting with a single conventional system using more reliable hardware to its various concepts of clustering. Other possibilities include a standby computer to temporarily record system transactions, with a later upload to the primary computer. This line of response also suggests data storage buffers if automatic data passing is the normal operation. The last resort is a completely manual control system such as paper and pencil. This manual system is also used extensively in bringing the system up and during normal system maintenance.

The important thing to remember is *the system will go down.* The ramifications of an unplanned system and/or component failure should be examined and alternatives should be put in the system if they can be justified.

Environmental Considerations

Computer systems generally are not finicky about their environments, but some site environmental considerations can exist. Computers, user terminals, sensing devices, and data entry equipment used in MES projects can be nearly any place within or out of the plant. Most applications will allow personnel and equipment to operate in normal environments, but in those systems where unusual conditions exist, it is likely that a product will fit. Manufacturers can provide information defining acceptable and ideal conditions.

Environmental concerns include temperature and humidity ranges, vibration loads, air quality, and electrical power requirements. Acceptable ambient temperature and humidity ranges continue to widen with each generation of hardware, and today most systems are content in temperature conditions considered intolerable for humans. Vibration loads should be considered. Mechanical devices such as disk drives and cable and backplane connections are subject to performance degradation and failure in high-vibration conditions. In these conditions, the equipment must be isolated with a suitable isolation or shock mounting. Air quality is a consideration where dusty or chemically corrosive conditions exist, as they will shorten the expected life span of practically any exposed electronic equipment. Particulate matter has the potential effect of altering electrical characteristics of some boards and also blocking cooling channels which provide needed air flow over hot components. Corrosive atmospheres do particularly nasty things to interconnect devices and media. The quality of the electrical power should be analyzed and must fall within manufacturers' recommendations. A steady and unfluctuating power source provides added mean time between failures. An uninterruptable power supply has long been a recognized necessity.

There are very few environmental conditions that limit the application of MES. Equipment or methods can be made to fit. The point is to give consideration to those conditions and address them accordingly.

Local Area Networks

Most MES use one or more local area networks (LANs) to connect multiple devices or systems in a limited geographical area. A LAN can consist of systems attached to a single length of cable, or the LAN can be formed by connecting a central system to each system in the building. A LAN is the combination of hardware and software used to control the interaction and transmission of data along and between systems.

LANs are used in manufacturing to connect devices such as robots, programmable controllers, controlled machinery, and sensing equipment such as bar code readers. Linking the office, information systems, engineering, and manufacturing provides the ability to communicate throughout the company and transmit data to any and all nodes on the network. An integrated environment allows different LAN technologies to communicate. PBX networks and PC LANs connected together or to other LANs provide a powerful network capable of serving as a corporate data network that can handle many environments and a diverse combination of data traffic.

The devices connected by the network are called nodes. The physical layout of the network is the topology—the way in which the nodes of the network are interconnected. The goal of topological design is to achieve a specified performance at minimal cost. LAN topologies can be divided into three basic types: star, ring, and bus.

In a star topology, all nodes join at a central node, creating a star-shaped configuration. The central node acts as a routing switch for arriving signals from each connected node. The network's performance is dependent upon the central node. Both advantages and disadvantages of this arrangement stem from this centralization. If the central node fails, the entire network fails, which is this topology's greatest disadvantage. However, the central node allows easier network monitoring and control. These networks are easier to service since most servicing is done from the central node. The star network often needs more cabling than other topologies and can be difficult to reconfigure.

Another type of network is the ring topology. In a ring topology, the nodes are connected in a circular pattern. Data must travel node to node around the ring. The sending node simply transmits its message toward its neighbor node in the ring. The message is then passed around the ring until it reaches the intended node. Each node needs only to recognize those messages intended for it. The ring topology eliminates the dependency upon a central node, but overall network reliability depends upon the reliability of each connection along the ring.

The simplest topology, a bus, is a linear run of cable with nodes connected along the entire length of the cable. Nodes are connected or tapped into the main LAN cable as required. Transmissions from a node flow in both directions toward the ends of the bus cable. The destined receiver must recognize data intended for it and read the message as it passes by. Unlike a ring topology, nodes in a bus do not retransmit messages. Long bus cables may require a repeater to regenerate the data signals.

The transmission medium is the physical connection over which communication between nodes takes place. It is sometimes called the link, line, or channel. Various types of media are used in networks. One common medium is twisted wire, originally developed for telephone use to transmit voice signals. Insulated pairs of wire are bundled, forming a cable. Twisted pair wire is low cost and easy to install but is more susceptible to electrical interference than other LAN media.

Coaxial cable is the most common physical medium for LANs. It combines robust construction, light weight, and modest cost with good electrical isolation, reasonable range, and capacity for high data rates. Coaxial cable consists of a central conductive wire surrounded by a shield of fine copper mesh and/or an extruded aluminum sleeve. A non-conductive material is between the shield and the center conductor. Coaxial cable shielding provides good immunity to noise interference. The increased wire insulation enables higher frequency transmission than twisted pairs because less energy is dissipated. Coaxial cable can be easily tapped for installing new nodes.

Fiber optics is another medium that has several advantages, including wide bandwidth, excellent noise immunity, ground isolation, high-voltage isolation, non-electronic radiation, small size, and light weight. Fiber optics use in bus topologies is limited because of the difficulty of making taps on fiber optic cables.

Transmission media interconnect all nodes in a network topology. However, for proper communication between nodes, data must be able to travel along any transmission path without colliding with other transmissions. Network access methods are the techniques by which nodes gain the use of the physical network medium to send a message across a network. These methods ensure that only one node transmits at a time on a channel, or, if more than one transmits, the proper recovery action is taken to provide correct data transmission.

Data Entry

Few computer systems can operate very effectively without information, and MES systems are no different. The adage "garbage in/garbage out" means the computer can be no more effective than the data being fed into the system. The issue here is how to input all necessary data and ensure data accuracy. The best condition is machine-to-machine communication, where data is read by a device and transferred to the system. Operator keypunch data entry is any system's weakest point. In a system with 99.9% accuracy that has 10,000 data items per shift (a small system), there would be 100 errors per shift. A system with 100 errors per shift will never be acceptable. System accuracy objectives must be in the 99.999% accuracy range or better.

The following definition of data collection is provided by Jonathan Cohen in his book *Automatic Identification and Data Collection Systems* (recommended reading):

Data collection is the recording of data items concerning an event or events, at the place where the event occurs and at the time of the event (or thereabouts).

His comments provide a number of reasons for automatic data collection:

- The timely availability of data
- The accuracy of data
- The importance of data quality

Data entry devices can range from manual data entry through a terminal keyboard to automatic entry through bar code readers, sensing devices (scanning or measuring), voice encoders, radio frequency transmitters, touch screens, programmable controllers, or other computer systems.

Bar Code Symbology

The most common form of data entry or data collection by automatic identification is through the use of bar codes attached to the item to be identified. A scanner is used to read the bar code label information which is then sent to a decoder for conversion to an electronic version of the data. Data collection through the use of this technology is very cost effective, provides very fast data entry, is very accurate, and is easy to use and understand by plant personnel.

There are four basic parts to a bar code system: the label, the scanner or reader, the decoder or data entry terminal, and the interface. A computer is usually also required to make the most effective use of the information that a bar code system generates.

There are two considerations for the bar code label: the production method and the bar code symbology used. Bar code labels can be purchased preprinted or printed at the point of use. The preferred production method depends on the intended use. Preprinted bar code labels can be purchased from suppliers, and equipment ranges from state-of-the-art photographic systems to offset printing and standard dot matrix or laser printers. There are several advantages to using preprinted labels:

- High-density bar code labels are available.
- Precise tolerances can be maintained.
- Labels can be laminated or otherwise protected.
- Labels can be produced on non-paper substrates including ceramics.
- A wide choice of adhesives and packaging is available.

There are numerous producers of preprinted labels. As with other commodities, quality and reliability may differ significantly between producers. A preprinted label is a purchased part and should be subject to quality control. The disadvantages of pre-printed labels are higher unit cost and the necessity to predetermine label content. Preprinted labels are usually higher quality than on-site label generation products and, in general, use less of the total tolerance of a bar code system.

Each bar code symbology has its own defined specification, calling out the ratios and combinations of bars and spaces used for encoding information. Three other considerations are:

- Print contrast (difference in reflectivity of the substrate and the individual bars)
- Freedom from voids and specks (unwanted gaps in the code or ink present in the space area)
- Compatibility with the equipment being used to read the label

Many different bar code symbologies are in use. Several general purpose and special purpose codes have been developed, all with relative strengths and weaknesses. The two symbologies discussed below are the most commonly used today in industrial applications.

Code 3 of 9—This symbology encodes a full uppercase alphabetic and numeric character set and several other characters in variable-length bar code symbols. The name "3 of 9" is derived from the character format where each contains three wide elements (bars and spaces) out of nine. Code 3 of 9 is highly immune to substitution errors (where one character reads as another), and it is well proven by extensive field experience. The highest standard printing density is 9.8 characters per inch.

Interleaved 2 of 5—This symbology encodes numeric-only data in a high-density format. The name "interleaved 2 of 5" is derived from the character format where pairs of digits, each encoded as two wide elements out of five, are interwoven, one represented in the space widths and the other in the bars. Interleaved 2 of 5 can encode any even number of digits but is prone to false reads. Full data security is achieved only if the reading device is programmed to accept messages of predetermined or fixed length, which may vary between applications. The highest standard printing density is eighteen characters per inch.

The bar code methods discussed above are limited in the amount of information that can be encoded. New symbologies to encode more information in a smaller space are rapidly emerging. These newer symbologies are usually referred to as 2D, with the labels using these symbologies sometimes referred to as a Portable Data File (PDF). One example of this technology is the PDF417 (see Figure 7.1), a two-dimensional, stacked bar code symbology* developed by Symbol Technologies, Inc.

Figure 7.1

3 of 9
15 Characters ...

PDF417
1456 Characters ...
(Gettysburg Address)

In PDF417, the basic data unit or minimum segment containing interpretable data is called a **codeword**. Every codeword in the symbol is the exact same physical length, and each codeword can

---

* The illustration of the two bar code technologies was supplied by Symbol Technologies, Inc.

be divided into seventeen equal modules. The size of one of these modules is the "**x-dimension**" of the bar code, or minimum element width. Within every codeword there are four bars and four spaces. The minimum number of modules of any bar or space is one; the maximum is six. In all cases, the four bars and four spaces of any codeword measure seventeen modules in total. Thus, the name PDF417 (or PDF "4-17") is derived from the structure of the symbology.

Many of these new symbologies have been developed for specific purposes, such as the MaxiCode developed by UPS. Many, however, are in the public domain. For further information on current bar code technology, including supporting equipment suppliers, contact the trade association: AIM USA, 634 Alpha Drive, Pittsburgh, PA 15238-2802.

Bar Code Label Scanners

The scanner serves as the eye of the bar code system by converting a visual image into an electrical signal. When choosing a scanner, the following should be considered. Do you need:

■ Analog or digital output?
■ Visible red or infrared illumination?
■ Contact or non-contact?
■ Fixed or moving beam?

The proper selection of a scanner depends on its compatibility with the application, the label, and the decoder. These choices apply to both hand-held and machine-mounted devices.

The choice between analog and digital output is primarily a matter of decoder compatibility. The choice of visible red or infrared illumination is governed by the type of ink used in printing the bar code label. Infrared illumination is more tolerant of dirt and contamination typically found in an industrial environment. Scanners using infrared illumination require absorptive inks. Visible red light is required where dye-based inks are used and for some laser-etched bar codes. The choice between contact and non-contact optics is determined primarily by use. In general, contact scanners offer lower cost, while non-contact devices offer greater depth of field and flexibility. The choice between fixed and moving beam scanners depends on application. Fixed beam reading is used in most bar code applications. A moving beam reader has definite applications where it is difficult or impossible to touch the bar code label.

The decoder is the brain of the bar code system. It takes the electrical signal from the scanner and converts the timing relationships of bars and spaces into an electronic representation of the information encoded. Factors to consider in selecting a decoder are:

- Scanner compatibility
- Output required
- Portability
- Need for display
- Need for key entry
- Number and variety of codes to be read
- Environmental factors

Scanners come in a variety of shapes and types:

- Hand-held wand
- Hand-held laser
- Fixed-mount moving beam
- Portable
- Hand-held scanner with computer and radio transmission abilities
- Radio frequency vehicle mounted

Scanners are available to meet nearly any application imaginable. Bar code application technology continues to evolve at a fast pace and is the most common method of entering data into an MES. Review your options with a number of suppliers to get the latest ideas for your system.

Bar Code Data Messaging

Beyond item identification, bar codes can also be used to send messages to the system. These messages can be whatever you want them to be. Figure 7.2 shows the use of a bar code menu to communicate directly with the MES from a workstation.* Each employee has a menu at his/her workstation with a number of different predefined messages. A wand scanner is used to scan the employee's ID badge bar code and a message from the menu.

Figure 7.2

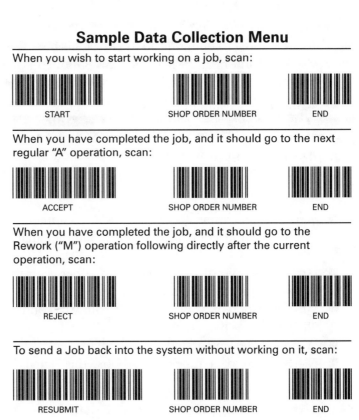

Sample Data Collection Menu

When you wish to start working on a job, scan:

START   SHOP ORDER NUMBER   END

When you have completed the job, and it should go to the next regular "A" operation, scan:

ACCEPT   SHOP ORDER NUMBER   END

When you have completed the job, and it should go to the Rework ("M") operation following directly after the current operation, scan:

REJECT   SHOP ORDER NUMBER   END

To send a Job back into the system without working on it, scan:

RESUBMIT   SHOP ORDER NUMBER   END

If you lose track of what you've scanned and wish to start over, scan:

---

* Bar code menu illustration furnished by I-O Solutions, Beaverton, Oregon.

This is a custom programming feature that can be easily added or revised as new ideas develop. The tool is very flexible and easy to use by everyone.

Radio Frequency Data Communication

A bar code data collection system can be extended by tying in radio frequency data communication (RFDC). There are two applications of RF in information management. One is RF identification, where a signal-emitting microchip is embedded in the product to be tracked. The chip emits a signal that uniquely identifies the article. The other is RFDC, where RF is used to transmit or send information through radio instead of over wire.

RFDC expands on a wire-tied data collection system by replacing the wires with RF signals. RFDC systems typically consist of the following:

- A controller connected to the host computer
- A base station connected to the controller that transmits to end devices
- An access point that acts as a bridge between the enterprise network and the radio system
- An RF reader or an RF network gateway connecting a non-RF reader to the network
- A transmitter integrated into each reader
- When needed, a repeater to boost the signals

The major advantage of RFDC is that wires and cables are eliminated, providing portability with instant access to the computer system. Most RF systems are set up for two-way communication with the computer system. There are other examples of RF in use, but the most common is probably send/receive units mounted on forklift trucks.

Radio Frequency Identification

Another method of item identification is radio frequency identification (RFID), where a transponder tag is attached to the item to be identified. Radio frequency identification is accomplished through high- or low-frequency radio signals providing data that can be read through virtually anything except metal. Most environmental conditions, including grime, dust, oil, wood, water, and other materials that often inhibit bar code applications, have no effect on performance. RFID provides a very accurate method of identifying items.

A system typically consist of four functional parts: (1) a passive tag or transponder, (2) an antenna, (3) a sensor, and (4) an interface. When a tag comes within range of an antenna transmitter, it receives the signal and automatically transmits its programmed data back to the antenna via a digitally encoded signal. The antenna receives the encoded signal, which is routed to the sensor. The sensor filters the signal and transmits it to the data collection computer via the interface.

The two main types of RFID tags in use today are passive and active. A passive tag reacts when prompted by the antenna interrogation, responding with an identifying message. Tags are available either reprogrammable or non-reprogrammable and typically provide ASCII alphanumeric data. Active tags carry their own independent power source, normally a long-life lithium battery. This enables the tag to receive and store data in real time as opposed to sending out a fixed signal. It also means that the signal range can be much greater. Unlike most other forms of auto-ID, the distance between the encoded data and the reader can be quite large. RF tag systems that operate on ultra-high frequency (UHF) can reach a distance of a few yards. (RF tags operating on low frequency rarely exceed one yard.)

When considering RFID systems, there are some items to examine:

- **Data Capacity**—The amount of data necessary for the application
- **Range**—Distance from the interrogator to the item to be read
- **Environment**—RF technology is susceptible to interference

RFID is a proven technology that has been successfully applied in many areas of manufacturing where a bar code label might not suffice; however, there are some limitations, including government-regulated use, and environmental considerations, such as electrical noise interference.

Voice Recognition

Another form of data entry is voice recognition, an excellent tool to collect data where a human interface is necessary and hands and eyes can be more effectively used in the operation. Systems typically consist of a voice terminal, a user-worn headset with a microphone and a lightweight belt radio, and an RF base station that receives the information from the wearer. The information is transferred to the data collection system by the base station. Most systems use artificial voice prompts and record the response to each prompt. A number of built-in data security methods (most systems recognize only the voice of trained users) help to provide highly accurate data. A broad range of application possibilities are available with voice recognition, including combining voice recognition with bar code scanning. The advantages of this method of data entry are:

- Leaves the employees' hands and eyes free.
- Provides immediate-use data entry.
- Allows attribute and numeric data to be collected.
- Accuracy of the data is excellent.

Vision Systems

Vision systems are another technology used where more conventional methods of identification are not possible. When vision systems are used in MES applications it is for purposes of item identification and data collection, the reading and data decoding of an identifying characteristic. A vision system consists of a camera that captures pictures of the identifier. These pictures are sent to a digitizer and on to a host computer. Vision systems considerations include:

- Ambient light including shadowing, sunlight, and glares can have an adverse effect.
- The item to be identified must be correctly positioned and stabilized.
- The identifying mark must be consistent.

Device Control Systems

One of the most common sources of data is the control systems of equipment on the plant floor, primarily programmable logic controllers (PLCs). A typical application is the download of machining instructions to a numerically controlled machining center (CNC), and the upload of data from that machine could include quality assurance information, run time information, or tool condition information. Many devices on the plant floor run by computer or PLC generate data to serve their internal needs and provide a buffer of information that can be retrieved through data collection and used for other applications within the MES. The programming requirements between existing plant floor systems, which can include SCADA systems, PLCs, and computers, are a normal part of MES implementation.

Touch Screens

When the MES must communicate directly with the plant floor employee, convenience is a major factor. Touch screen terminals provide excellent convenience through their ability to input data without a keyboard or a scanner. This tool is often used when a terminal is placed at the plant workstation to provide information or retrieve information specific to that position. By posing questions on the screen with suggested answers, the operator is required to merely touch the screen at the correct indicated position. The terminal can accept the response and process the information to the host computer.

Other Computer Systems

Other computer systems in the plant, particularly the support functions discussed in Chapter 5, are excellent sources of information. Examples include tying in a time and attendance system to gain access to employee information or a connection to the product data management system allowing retrieval of product information for use by a machine tool or a coordinate measuring machine. Until technology progresses to the point where "plug and play" is common, each of these systems must be brought together through a programming interface or some kind of data transfer connection. This is not particularly difficult if the software on the intended data source can be examined and revised. Armed with the correct knowledge of the existing software, the interface, along with the data exchange requirements, can be defined and programmed to provide the appearance of a seamless system.

Data is the obvious lifeblood of any computer system, but it is even more so with on-line systems that depend on the latest information. This applies directly to MES, as it is the generation and management of data that will cause subsequent events to occur. It is important that the system requirements focus on the data and its uses. With a well-defined system requirements definition document, it becomes much easier to make decisions on equipment issues. Many industrial buildings are designed and then the process is made to fit the building. It is unlikely that anyone might willingly agree with this method, but it happens. The MES is similar in that the application requirements (manufacturing's needs) must come first, and they must drive the thinking and the process of equipment selection, not the other way around. Much of the technical consideration should lie with the vendor, but it is important to adopt a risk-averse strategy—that is, to avoid the risk of system failure through the development of a good requirements specification.

References

AIM (Trade Association of the Automatic Data Collection Industry), 634 Alpha Drive, Pittsburgh, PA 15238.

Cohen, Jonathan, *Automatic Identification and Data Collection Systems,* McGraw-Hill, London, 1994.

CompuSpeak Laboratories, *Voice Recognition,* CompuSpeak Laboratories, Olathe, KS.

IBM Corporation, *IBM Strategies for High Availability,* IBM Corporation, Armonk, NY.

Intermec Corporation, *Understanding Radio Frequency Data Communication,* Intermec Corporation, Everett, WA.

Itkin, Stuart and Josephine Martell, *A PDF417 Primer, A Guide to Understanding Second Generation Bar Codes and Portable Data Files,* Symbol Technologies, Bohemia, NY, 1992.

Longacre, Andrew, *Emerging Bar Code Technologies, Technical Review 1996,* AIM International, Pittsburgh, PA, 1996.

Telsor Corporation, *Radio Frequency Identification,* Telsor Corporation, Englewood, CO.

Welch Allyn, Inc., *Bar Code Handbook,* Welch Allyn, Inc., Skaneateles Falls, NY.

8 The MES Implementation Team

MES has been defined as a computerized integrated system to execute the production plan, but thinking about the potential information suppliers and users of the system should not be limited; it should be inclusive. When designing information systems, it seems like everyone has thoughts on what could be or should have been done. Designing an integrated MES will, sooner or later, require broad input from within the organization that cannot be satisfied by any vendor. Using an implementation team with input from many sources will provide a better system, greatly smooth the way for system acceptance, reduce the risk of system failure, and ensure that the system addresses the needs of the organization instead of just those of an individual department.

There are a number of reasons to consider who might be a good contributor for the implementation team. The assignment is critical and crucial to a successful system. The primary responsibilities of the team begin with defining the requirements of the system, selecting the vendor or vendors, and implementation guidance when the system goes on-line. MES is a process that crosses boundary lines between business functions such as manufacturing, information processing, purchasing, and engineering; between departments such as production scheduling and production supervision; and often between plants where there are multiple locations. Developing the system requirements and getting user acceptance can become difficult if the need for a broad perspective is not recognized early in the process. Selection of vendors will be a significant task, even if a company has standards for equipment, operating systems, and database systems. Implementation will require some continuing involvement, not only to monitor progress but also to help sell the changes people will see.

The following team responsibilities and formation were outlined in a recent paper published by MESA International and have been revised some by this author. The paper suggested this team be broken into two groups: (1) the oversight or review team and (2) the evaluation/selection team. The two groups are differentiated by number of people, areas of expertise, responsibilities, and time commitment as follows:

MES Oversight Team

- Four to six people representing various manufacturing functional areas including purchasing, materials management, and scheduling, along with a production foreman or equivalent
- One person from manufacturing leadership (plant manager, production supervisor, etc.)

Responsibility

- Meetings two to four hours, one time per week
- Review progress of selection team
- Assure that the evaluation/selection is being followed
- Provide functional area expertise as needed
- Approve the final decision

MES Selection Team

- Three to four full-time representatives:
  (1) Information systems
  (1) Production (general supervisor or equivalent)
  (1) Production control or scheduling
  (1) Quality assurance or process engineering
  (1) Purchasing/materials management
- One outside consultant/systems supplier: one to two days per week
- One full-time project leader with many of the following qualifications:

☐ Must be thoroughly familiar with company operations and accepted within top management

☐ Must have knowledge of entire plant floor operations with experience in more than one functional area

☐ Must understand and be an innovator of manufacturing methods

☐ Must understand the needs of data users and the distribution of data within the company

☐ Must understand MRPII, scheduling, inventory, material movement, quality assurance, maintenance, etc. and their interrelationships

☐ Must understand and support managerial strategic concepts relating to manufacturing

☐ Must be a good team leader able to guide discussions toward a consensus and a conclusion

☐ Is responsible for arranging the team's meetings, trips, presentations

☐ Acts as a single point of contact for vendors

It is not necessary that this person have in-depth knowledge of computers. This position should not be involved in such issues as database systems, operating systems, or bits and bytes of computer or communication equipment.

Responsibilities

■ Determine the high-level system requirements.

■ Create an initial vendor list.

■ Review/evaluate products and survey vendors.

■ Prepare the requirements definition document.

■ Meet with vendors.

■ Compile the results and recommend the final product/vendor.

The size of the team and the amount of time that must be dedicated to the project are based on the size and breadth of the project.

There are many sources of information for applying and purchasing MES; they include:

- Trade and professional association shows (APICS, MESA, SME, etc.)
- Trade publications and professional journals (APICS, MESA, SME, *Managing Automation, Manufacturing Systems, Controls Engineering, Industrial Engineering,* etc.)
- *MES Directory and Comparison Guide* (Thomas Publishing)
- Manufacturing consultant publications (AMR and Gartner Group)
- Referrals from manufacturers in your or similar industries

Once the decision has been made to implement a system, it is suggested the company adopt a strategy of being **risk averse**—that is, to avoid the risk of system failure. To be averse is to be unwilling, disliking, or have an aversion to something, in this case, risk. The most important aspect of being risk averse is to accept system responsibility—define and understand in detail what you want the system to do. This requires more up-front work, but the payoff will be very rewarding. As you define what you want the system to do, you must think through how you want to manage your business—a responsibility that should not be transferred to a vendor. This is not to suggest not using the many excellent resources that are available. All resources, ranging from suppliers to consultants, should be considered, but in the end the selection team must take responsibility for their application and the results.

The initial task of the team is to develop a system requirements definition. The requirements definition should define and analyze the current problems and opportunities, the objectives and goals of the system, and the expected results of the project.

- Identify the fundamental problem or problems the system is to address (the "as is" condition).
- List specific needs the solution must satisfy related to functions and tasks (the "to be" vision).
- Describe the physical system environment, the expected users, and their computer skill level.
- Indicate any hardware and software preferences or needs.
- Describe minimal quality, performance, security, and support requirements.

- Describe any compatibility and migration needs, including existing systems, their capabilities, and their new requirements.
- Describe the system support-related needs.

The MESA International paper "MES Software Evaluation/Selection" suggests the requirements definition document contain: (1) a description of the business and the areas to be addressed, (2) a detailed list of functional and technical requirements, and (3) a representation of events that occur in the shop floor environment and the responses that the system is expected to generate. In addition, most companies require an economic justification of the investment prior to purchase that ties the cost to direct financial returns. It is safe to say that preparing the requirements definition is arduous and detailed, but a significant number of critiques have been written about the pitfalls of computer system applications, and most problems can be connected to incorrect assumptions regarding the details on what the system is supposed to do. The correct foundation must be developed at this point with and by the selection committee. This is a major part of **risk-averse strategy**.

Some companies will have a beginning for their system from internally generated requirements. If not, exploring the following list of items might provide some thought-provoking ideas:

I. **Manufacturing Control System Review**
   A. Inputs
   B. Outputs
   C. Reliability/usability
   D. Who are the users? What complaints exist?
   E. Timeliness and frequency of input and output
   F. Where and how can the MES connect?
   G. Review available product data—routings, standards, accuracy, etc.
   H. What inventory data is available and how accurate is the data?
   I. What connections or outstanding issues exist between departments?
      1. Purchasing and receiving

    2. Engineering and CAD/CAM connections to the plant floor

    3. Sales department and knowledge of finished goods inventory

    4. Accounting and the source and accuracy of cost and labor data

  J. Examine in detail the shop floor control module and information flow

    1. Outputs

    2. Inputs

    3. Timeliness

    4. Batch/on-line

    5. Data availability

  K. What weaknesses are seen by the users?

II. **Shop Floor Environment Review**

  A. Review the workstation and/or work cell physical layout and logic/flow connections

  B. Which workstations can be integrated into the MES?

    1. Machine control integration

    2. Personnel input integration

    3. Physical and logical connection to inventory

    4. Relationship to material handling equipment

      a. Manual

      b. Automatic

  C. How does supervision tie to a workstation?

  D. How does the workstation tie into the overall management system?

  E. Review the current and/or possible workstation/work cell relationship

  F. How many workstations and work cells exist and how are their operations currently identified? (List all possible operations required and possible for each workstation)

  G. Review current scheduling methods for shop orders, work cells, and workstations

  H. What time frames are or can be allowed for workstation reporting? Report once per day, after each piece, after each order, etc.?

I. What scheduling and other workstation decisions can be locally controlled versus MES controlled?

J. What data arrives with the schedule and in what format?

K. Review an example of routings and alternate routings

L. Review a sample of all variance reports, labor, quality control, material, etc.

M. Review samples of work orders, with an explanation of all codes and numbering systems

N. What is the procedure to change an order schedule or priority?

O. What is the procedure for engineering change orders?

P. What is the procedure for lot splitting or combining?

Q. What is the procedure to change lot or order quantities?

R. Review current inventory methods, equipment, and control procedures
   1. Storage algorithms
   2. Data requirements
   3. Inputs and outputs
   4. Receiving department quality assurance methods and data requirements

S. Review in-process costing and pricing methods

T. If bar codes are used, what information is in the bar code and who are the users?

U. What are the rules and procedures regarding quality assurance?
   1. Tracking requirements
   2. Variance report requirements
   3. Report and data distribution

V. Review current and/or planned labor data collection requirements including time and attendance, piecework considerations, quality assurance, set-up, machine downtime, etc.

The next step after the requirements definition is the development of a functional design specification that describes in detail what the system will look like to the user. The functional design specification is done either by a vendor or the user and includes each screen, data field, keystroke definition, and all interfaces to external participants or data points. A functional design specification:

- Identifies all data input and output points and the data characteristics.
- Lays out each screen or man–machine interface with exact definitions for each screen, data field, and keystroke.
- Identifies and describes in detail any and all expert rules.

This effort may seem exhaustive and overly detailed, but these definitions and decisions must be made either by the vendor or by you, the buyer. If you leave these decisions up to the vendor, then you must run your business the vendor's way. **Can you take that risk?**

The rigorous detail suggested may not always be necessary. In many cases, standard off-the-shelf, fully contained systems may fit your needs. Even then, you should know in detail what is being implemented and accept responsibility for knowing that the system being purchased will address your business needs correctly. Unsuccessful systems are very costly in terms of both the effects on production and the negative public relations that will affect future projects. Here again, the suggested strategy is to be risk averse. Recognize the risk issues and deal with them up front before trouble begins. A major step is to accept system responsibility. Know what you are buying and know how you will measure the system's contribution to your entire operation, including the return on investment.

References

Charette, Robert N., *Software Engineering Environments, Concepts and Technology,* Intertext Publications, McGraw-Hill, New York, NY, 1986.
MES Software Evaluation/Selection, MESA International, Pittsburgh, PA, 1996.
Whitten, N., *Managing Software Development Projects,* John Wiley & Sons, New York, NY, 1995.

9 Supplier Evaluation and Selection

S upplier selection is a very important part of the MES implementation process. It is particularly significant in keeping with the **risk-averse** strategy. In the current issue of the *Thomas Register Directory of MES Suppliers,* there are dozens of companies listed, each with products and/or services that address MES opportunities. Choosing which of these companies to consider as a supplier requires an extensive evaluation to determine who could best satisfy the system requirements as they are currently defined as well as supply a system that can grow and change as the business changes. The review and selection of potential suppliers should begin soon after the implementation team has given a reasonable indication of the size, scope, and schedule of the project. An invitation can then be issued to a group of potential suppliers, requesting a response that indicates interest and specific strengths available to provide the best solution to the requirements definition. Their responses should also include a list of similar completed projects that could be contacted or visited. Suppliers are usually very willing and able to provide input and assistance during the definition phase, including site visits to companies where successful systems have been installed. Site visits can give the implementation team fresh information, especially what not to do.

There are many companies serving the MES market. Some are distributors of software and do very little, if any, software production. Some provide software only, leaving the installation up to the user. Other companies will provide complete turnkey systems ranging from reengineering to training and system support. Other companies are interested in providing only narrow niche solutions that may be stand-alone or designed to fit into a larger system. Still other companies may offer a combination of all of these, buying software and hardware from many sources and knitting them all together with their in-house-developed software. Some suppliers have stronger skills in system development or plant floor operations or in control system installation. With the wide variety of companies serving this market, searching for the few companies to participate in your project is a significant task, one that will not be accomplished overnight. To make this selection process easier to understand, potential suppliers can be divided into four categories:

- **System Integrators**—System integrators are software, hardware, and services providers that combine products from various suppliers into an integrated system. These companies may or may not provide programming to knit the various products together. System integrators might be thought of as general contractors, with most components of the system provided by other companies, their subcontractors. The following are common characteristics of systems integrators:
  - ☐ They are frequently distributors or dealers for many of the products that might be included in the system.
  - ☐ Products or services can be from within the company or procured from outside sources.
  - ☐ These companies are more likely to be local, a real advantage.
  - ☐ They often have broader experience across a variety of installations.
  - ☐ Their major strengths usually lie in the areas of installation and local service, with weaknesses, if any, most likely in technical software and system application knowledge.
- **Point Solution Vendors**—Point solution vendors provide software products that address a single or narrow issue. These products are usually operated as stand-alone systems. An example is a company that supplies a statistical process

control system or a time and attendance package. These vendors' characteristics include the following:

☐ Most applications are aimed at certain specific applications such as scheduling, maintenance, statistical process control, etc. The broader view of manufacturing may not be adequately addressed.

☐ These products are frequently aimed at a very defined user market.

☐ Modification of the system functions and screens is very limited or not possible. What you see is what you get, and although the product may be done very well, cross-industry applications may not always fit.

☐ The fact that the system is stand-alone could be a good thing, but is it possible to integrate the package into the larger MES objective? Can data be retrieved for other uses?

■ **System Developers**—System developers are companies that develop broad MES software systems that can be reused and/or modified to fit new applications. These companies can usually provide an array of services, such as system design, software, hardware, and integration, including installation and wiring. This type of supplier might be thought of as a general contractor that develops the most critical component, the application software. System developers usually fit the following descriptors:

☐ The software can be customized to fit your application using previously developed programming code.

☐ The capability of the supplier is frequently built on a specific industry or application knowledge base that is expanded to fit new applications.

☐ The supplier can usually provide a complete, installed system and accept system responsibility as part of the contract.

■ **System Consultants**—Some consulting companies specialize in MES, providing advisory services separate from any hardware or software purchases. Consultants can provide an independent third-party view, along with time and expertise that may not be available elsewhere, either in your company or from a vendor. Their services can assist in many ways, including simple presentations on MES, assisting or developing the requirements definition, evaluating vendors, developing an implementation plan, and training. Consulting companies come in all sizes and can participate as a part-time member of your team, providing input on specific issues, or as a full-time team member, supplying general guidance and counsel.

Suppliers will vary, but most can furnish a complete system, including software, computer and communications hardware, installation, wiring, data collection devices, database products, training and operating manuals, training programs, and device programming. There will always be a large amount of overlap between these definitions. The intent is not to limit any supplier but rather to give the buyer some insight into how various companies address this large and varied market.

Any vendor should be able to provide the following items of information on its product and the company:

- Detailed product application information
- A list of current users that you can contact and possibly visit
- A documented training program
- A system for software upgrades
- System software documentation
- System licensing information
- On-line and/or telephone system support

The following items must be considered to help define the list of possible vendors for your project, particularly if the application is large and complex.

- **Financial Condition**—MES can involve large financial commitments. A review of recent and current financial statements of proposed vendors is suggested.
- **Personnel**—A proposed project manning chart along with a resume for each person with managerial responsibility will help assess capability.

- **Project Management**—A proposed project management plan is necessary, complete with milestone review dates and progress-measuring criteria. Large projects might include project scheduling techniques such as critical path, with monthly updates and progress payments tied to meeting project milestones. The definition of milestones and deliverables is very important.

- **Operating Information Security**—Even if a requirements specification has been developed for the MES, any supplier must have detailed information on how your company operates. This might include new product development plans, production rates, and other information that should not be made public. Most vendors will recognize their responsibility and act accordingly, but this aspect should be considered, especially when the vendor you have in mind has in-depth experience in your industry. That experience is very helpful, but it must be measured against possible exposure. Non-disclosure agreements are commonly used throughout the industry.

- **Software Ownership, Changes, and Licensing**—Most companies that develop and provide software for use by others do so on a licensing basis. For a fee, the user company is allowed to use a copy or copies of identified computer programs. Although this is industry practice, consideration should be given to programs developed for a specific project and paid for by the user. Another issue to be considered is changes. Who can make changes? Who owns the changes after they are made? Will there be future system upgrades, and how will they be priced? If the supplier goes out of business, will the source code be available? There are no absolute rules that cover this except in the contract between the buyer and the seller. It is up to both parties to review and understand these issues.

- **Company Standardization Programs**—Many companies have developed standards for all information systems; they usually apply to equipment, computer operating systems, and database management systems. Most MES suppliers recognize the importance of these standards and can comply.

- **Manufacturing Execution Systems Are Not a One-Time Thing—** It is very important to recognize that the implementation of an MES is not an event but is the next step in the production improvement process that is likely to continue for years into the future. The definition of MES is the computerization and integration of your production process. When is the last time a change of any kind was made to that process? The requirements for change to your process will not decrease with an installed MES. If anything, expect more change suggestions (system improvements) as more people understand the process. The relationship between the company and the supplier should be considered in the long-range sense. **Changes and enhancements will be required**. A true partnering relationship based on trust and professional performance is very desirable and can provide rewards for both parties. The MES system will evolve, and it will be much easier to work with a supplier that has knowledge and experience in your plant than it would be to start over again with a new supplier.

Once the final list of potential suppliers has been established, the request for a proposal along with the specification information is distributed. It is difficult to generalize how long proposal preparation will take, but four to six weeks is not long for a complex project. Another four to six weeks will be required to analyze the proposals, make a decision, and negotiate the final contract. The contract should include a schedule with specific milestones and well-identified deliverables. Depending on the complexity of the system, delivery could be as short as a few weeks or as long as two years.

It is at the time of deliverables that projects can falter and the relationship between the supplier and the customer can be most vulnerable. As the supplier is proceeding with its plan, it will provide evidence of progress and seek concurrence on issues that arise, frequently offering new ideas. Notwithstanding the problems of a deficient system, there is nothing that has a greater negative impact on an MES project than **change** or **delay** caused by waiting for finalization of outstanding issues. The risk-averse strategy is a very important ally, and a well-done requirements definition is the only safe answer. Changes and delays cost money and time and can adversely affect the focus of both parties. Unless there are genuine system misapplications, examine whether the best action is to continue with the project as originally contracted to completion and make the necessary changes as a system revision.

The changes just mentioned are one kind of change. Another kind of change stems from the constant evolution that is part of every business. As indicated earlier, a manufacturing execution system should be seen as a continuing process; suggestions of system enhancements are to be expected, with changes and new ideas the norm. The objective is to manage these changes in ways that are cost effective and deliberate, not reactive and expensive. If there is any question about potential change, think about where your system will be in five, ten, or fifteen years.

- **Company processes change**—How will the MES be affected by revised manufacturing practices or process changes? The MES cannot be a controlling factor that prevents change. The system must be adaptable to whatever change might develop.
- **Technologies change**—If the past is any indication, the information and applications available to each user of information technology will be faster and more effective. The technology that will be available in ten years and how it can be integrated into the MES without disruption or starting over is an important consideration.

Change should be expected as a normal part of the process. Plan for change and act accordingly—during the implementation phase and after the system is on-line.

Selecting the supplier for your MES can be similar to selecting a marriage partner. The relationship will be intimate, must be mutually interdependent, and is likely (hopefully) to be long term. To take the analogy one step further, separations and divorces will be very difficult and costly. Investing the time and taking a few organized, objective steps up front will help ensure a selection that will go a long way toward accomplishing your **risk-averse** strategy.

10 Scheduling, Simulation, and Constraints Management

I f economics were not an issue, management of the process of manufacturing would be fairly simple. When economics is part of the equation (which is nearly always the case), there are constraints against the use of resources in the form of inventory levels, investment in equipment, necessary production sequences, the number of people available to respond to a current event, and competitive market costs. Constraints are usually addressed through scheduling the available resources as effectively as possible, a task that is rarely straightforward and, in most cases, is very complex.

Scheduling, in the sense of specific timed actions, is usually not a part of the planning layer. The planning layer is most often limited to assigning orders to time periods, often referred to as time buckets. Some systems can measure the quantity of work in a time bucket and indicate the level of capacity that is required in a given time frame, but this is frequently only a leveling function. The scheduling required on the plant floor must address the constant question of what must be done *next* to optimize resource utilization within the immediately current constraints of the organization and management goals.

Most manufacturing facilities use manual methods for production scheduling. Manual scheduling is very difficult due to possible combinations, current available resources, and the time required to look at the many possibilities. Scheduling is a very dynamic process; however, most methods are applied to a static environment. In order to make scheduling more responsive to changes of orders, equipment problems, and material shortage, it should be done online. This requires feedback information from the shop floor in order to adapt the schedule to the current factory condition. Newer computer-driven ideas on the market can accomplish this, and they provide real advantages over manual systems. They are constraints management and computer simulation modeling. Each has its area of application as a component of the manufacturing execution system.

In Chapter 4, the core functions of order management and workstation management suggested scheduling is a part of the core functionality because scheduling bears directly on orders and the use of resources. Scheduling is discussed in this book because of the importance of the subject *within* the manufacturing execution system and in order to show the newer sophistication of these tools as a part of the MES.

Scheduling means many different things to many people, especially in the world of manufacturing computer systems.

- **Capacity Planning**—The capacity planning module of most planning systems is a very basic form of defining what the plant can do in the theoretical and perfect world, based primarily on simple rules of each machine's capacity that measure a "rate" of output of the production system, usually measured in hours. For example, if 100 people in a plant work 40 hours each, the plant has a capacity of 4,000 hours of production. By comparing the current backlog of work in the master schedule to the 4,000 hours of "capacity," we can determine whether or not the plant's capacity is oversold. This is often referred to as infinite capacity planning, and the assumption is that the shop floor can react to the current capacity requirement by increasing capacity (working overtime) or decreasing capacity (temporary layoffs).

- **Finite Scheduling**—Finite scheduling is a more specific method of determining the production schedule by loading work into each available workstation or process. This method examines each order routing and uses operations and time standards to load into each workstation. The procedure usually schedules operations forward in time and recognizes that capacity is limited at any point in time and at any work center.

- **Forward Scheduling**—This scheduling method begins with a known start date and computes the completion date, usually proceeding from the first date to the last. Dates generated are generally the earliest start dates for each operation.
- **Backward Scheduling**—Backward scheduling is similar to forward scheduling except it begins with the planned completion date and computes the necessary start date for each operation.
- **Line Balancing**—This is the balancing of the assignment of tasks to workstations in a manner that minimizes the number of workstations and the total amount of idle time at all workstations for a given output level.
- **Sequencing**—Sequencing is a component of scheduling that acts on individual work orders or the queue of orders at a specific workstation. These sequencing rules are taken from *Manufacturing Planning and Control Systems.*
  - ☐ *Random*—Pick any job in the queue with equal probability.
  - ☐ *First Come/First Served*—Schedule the work in the order that it has arrived.
  - ☐ *Shortest Processing Time*—This has the effect of processing the smaller orders more quickly.
  - ☐ *Earliest Due Date*—If meeting the schedule without regard for other cost issues is the most important criterion, it can be argued that this method is fair.
  - ☐ *Critical Ratio*—This commonly used technique is calculated using (due date – current date) ÷ (lead time remaining).
  - ☐ *Least Work Remaining*—This rule is an extension of shortest processing time in that it considers all of the processing time remaining through completion.
  - ☐ *Fewest Operations Remaining*—This considers the number of successive operations.

- ❑ **Slack Time**—This variant of earliest due date subtracts the sum of set-up and processing times from the time remaining until the due date. The resulting value is called "slack." Jobs are run in order of the smallest amount of slack.
- ❑ **Slack Time per Operation**—This variant of slack time divides the slack time by the number of remaining operations, again sequencing jobs in order of the smallest value first.
- ❑ **Next Queue**—This rule considers the queues at each of the succeeding work centers to which jobs will go and selects the job for processing that is going to the smallest queue.
- ❑ **Least Set-up**—This rule maximizes machine utilization by recognizing the impact of set-up time on each machine.

This is not an exhaustive list. There are variations, combinations, and other rules for sequencing. Considering current conditions and being flexible enough to meet agile requirements will ensure the best fit of rules to your product and facility or specific workstation.

Scheduling is a very complex part of managing any manufacturing facility, not because scheduling rules are particularly complex but because the variables that go into determining a dependable schedule can be vast. Those variables include the initial planning system output, inventory availability, tool availability, and personnel skill availability, among many others.

One major contributor to modern manufacturing scheduling concepts is Theory of Constraints (TOC), developed by Eliyahu Goldratt in his books *The Race,* published in 1986, and *The Haystack Syndrome,* published in 1990. TOC is a different approach to scheduling based on the simple idea of addressing the constraint or constraints to otherwise successful production execution. Much in the same way that the weakest link is the breaking point in a chain, so too is a constraint the limiting point in a production plan. TOC deals with identifying and resolving constraints through the following five steps:

1. Identify the system's constraints.
2. Decide how to exploit the system's constraints.

3. Subordinate everything else to the above decision.
4. Elevate the system's constraints.
5. If, in the previous steps, a constraint has been broken, go back to step one.

Because the constraint or constraints set the pace for the plant (the system cannot operate faster than the slowest bottleneck), they are called **drums**, the pacesetters. To provide protection against disruption at the drum, a time **buffer** is provided. To ensure that inventory will not grow beyond the necessary time buffer, the rate at which material is released into the plant is limited by the rate of the capacity constraint. This is called the **rope**. This method of developing a synchronized manufacturing plan is thus often referred to as the **drum-buffer-rope** system.

The result of this technique moves toward a schedule based on real rules, including:

- The capacity constraint resource should dictate the schedule based on market demand and its own potential.
- The schedule for succeeding operations (including assembly) should be derived accordingly.
- The schedule of preceding operations should support the time buffer and thus be derived backward in time from the capacity constraint resource schedule.

A major part of constraint theory is the use of inventory to provide a buffer to protect critical schedules against disruption. The idea of inventory to provide insurance against unknown conditions is not new, of course, but this time the rules are more specific wherever they are applied:

- Concentrate protection not at the origin of a disturbance but before critical operations.
- Inventory of the right parts in the right quantities at the right times in front of the right operations gives high protection.
- Inventory anywhere else is destructive.

Mr. Goldratt's ideas are not academic concepts but instead came from those ever-present manufacturing-world issues of balancing and optimizing inventory, throughput, and operating expenses. His books are directly related to manufacturing issues, presenting a suggested paradigm change in business measurement methods. In *The Race,* there are some suggested rules regarding manufacturing management:

1. Balance flow, not capacity.
2. The level of utilization of a non-bottleneck is not determined by its own potential but by some other constraint in the system.
3. Utilization and activation of a resource are not synonymous.
4. An hour lost at a bottleneck is an hour lost for the total system.
5. An hour saved at a non-bottleneck is just a mirage.
6. Bottlenecks govern both throughput and inventories.
7. The transfer batch may not and many times should not be equal to the process batch.
8. The process batch should be variable, not fixed.
9. Schedules should be established by looking at all of the constraints simultaneously. Lead times are the result of a schedule and cannot be predetermined.

A number of suppliers of software systems incorporate TOC ideas. One supplier offers its systems in the following modules:

- **Production Planning—**
  - ☐ Rough-cut capacity planning
  - ☐ Material requirements analysis
  - ☐ Master planning
  - ☐ Reports related to the above issues
- **TOC Scheduling**—This module provides tools to manage the finite capacity based on TOC principles. Using this module you can:
  - ☐ Identify the constraints.
  - ☐ Exploit the constraints by scheduling them properly.
  - ☐ Subordinate the non-constraints to the schedule of the constraints and customer orders.
  - ☐ Identify material constraints.

☐ Generate material release schedules.

☐ Schedule incrementally so that new orders can be scheduled without disrupting existing schedules.

- **Production Control**—An integral part of TOC is buffers. Once schedules have been generated, the buffers need to be monitored to stay on schedule. Production control lets the user monitor the buffers and provides warnings about potential slips, allowing proactive action to stay on schedule.

- **Business Impact Analysis**—This module provides the ability to analyze capacity so as to generate maximum financial throughput.

- **Due Date Quotation**—This module allows the system to quote due dates for new orders that are received since the last schedule session.

As with most manufacturing software products, there is room for modifying these systems to fit your environment.

Another approach to scheduling is the use of simulation through computer modeling, where a logical operational model of the manufacturing system is created to provide a graphical finite-capacity planning and scheduling system. This system sees the world as a number of *stations*. Stations can be machines, workbenches, assembly positions, or any location where work is performed on a product. A group of stations that perform interchangeable work is called a *family*. Every station belongs to a family, even if the family only contains a single station. Families share a common input queue and work list for parts waiting to be serviced by one of the stations in the family. Each station can have one or more *calendars* associated with it. Calendars specify when stations are available for work.

Lots flow between families in the model. They consist of a quantity of pieces of a given part and flow according to a routing that is defined. A routing consists of a number of steps and indicates the parameters for that lot, such as the station, family, the setup, processing time required, and the operator class. These parameters define the operation performed on the part.

As lots move through the steps in the routing, they enter the family work list and queue for the family designated for that step. Idle, available stations in the family execute the *task selection rules* associated with that particular station. Task selection rules are the criteria used to determine which lot to work on next at any given station. The task selection rule allows the station to either pick a lot from the potential parts or wait for a better choice. The status of operators, tools, components, and other constraints may be considered in a task selection rule.

As lots are *simulated* through their routing, a schedule file is updated. This file logs the simulated time at which the operations take place. In addition, factory performance statistics are collected, indicating the efficiency expected if the factory is operated according to schedule.

This approach to scheduling rules makes it possible for plant scheduling personnel to define rules for each resource. These rules are easy to construct and are not limited to a single rule criterion or a single view of a resource. A rule is a series of filters into which potential tasks are fed. Each filter is a criterion, or test, that the lots must pass. The filters screen out more and more lots until either one or no lots remain for final selection. The rules can utilize decision trees, sorting, and filtering, and each rule can contain as many filters as necessary.

Scheduling Rule Example

Suppose you must develop a scheduling task rule for a station where the number of set-up changeovers must be minimized to improve quality or to reduce lengthy set-up time. When it is necessary to change to a new set-up, the station should not select a lot or a batch of lots that requires a setup being used by any other station in that family. If there is more than one lot or batch that meets the criteria, the one with the earliest due date should be chosen.

Once the goals and objectives to quantify schedule performance have been established, a decision can be made on how to develop scheduling rules. Rules should be developed off-line with information that is representative of the real scheduling problem. It is not usually effective for each resource to try to optimize its own utilization. It is better to coordinate all the machines, workstations, and resources in a system to work in concert. This is called ***coordinated rule-based scheduling***, and it requires the following steps:

1. **Identify the critical resources in the factory**—The critical resources are the ones that increase the overall throughput of the facility when the constraints are relaxed. Analyze production requirements (loading) to determine the resource with the greatest load.

2. **Keep critical resources busy**—Do this by keeping manageable levels of work in the resource queues, choosing tasks that minimize set-ups, and avoiding unnecessary set-ups. If necessary, critical resources should look upstream in an attempt to wait a reasonable amount of time for a better choice, rather than taking what is available in the current queue. Resources should also look downstream in an attempt to keep products flowing through the plant by choosing jobs that won't get bogged down in a large downstream queue.

3. **Identify server resources**—Server resources have low to moderate loads and can be used to aid critical resources. Server resources can ensure that the critical resources have work and that the work minimizes set-up time. Server resources should look downstream and feed the most productive work to the critical resources.

4. **Run the model and analyze the results**—Identify the schedule performance measures. If you see a trend that a rule change could possibly help, enhance the rule and run the model again.

5. **Repeat the previous steps** until you are satisfied with the schedule performance.

6. **Employ the selected rule** with "live" data to schedule the facility.

There are two ways to apply this system of rule-based scheduling to the manufacturing scheduling problem:

- **Preplanned Scheduling**—With preplanned scheduling, a finite capacity schedule is developed from an accurate status of the shop floor and then provided to the shop floor for implementation. If unforeseen events occur, a new schedule must be prepared.

- **Real-Time Scheduling**—With this system, the simulation is run parallel with real time. This requires the model to receive information as events occur on the shop floor. The model updates the status of the internal data structures as it receives the event messages.

The system provides two types of output from the simulation/ scheduling run: graphical and statistical.

- **Graphical output**—The user can create business graphs to track any statistic with a time line graph, bar chart, or pie chart. These graphs/charts are updated as the model progresses. The system creates an interactive Gantt chart that provides the simulation history for each resource and order. The user can select an event on the Gantt chart and get detailed information about that event, such as the quantity of orders in that station's queue, when an order was selected, the quantity of orders in the next step's queue, etc.
- **Statistical/history report output**—This includes the master schedule file, performance report, station report, and any other user-defined reports.

Factory performance measures such as queue statistics, station utilization percentages, current work-in-process inventory levels, on-time delivery performance, etc. are updated during the simulation and can be displayed graphically or in printed form, during and after the run. They are also summarized in a final report. The user can specify what statistics appear in the reports.

The material in this chapter probably has given you much more information regarding scheduling than you may have expected or even wanted. Scheduling is a vast subject, and specifics are difficult to explain without a real-world subject to examine. There are, however, common items addressed in any schedule. Any concept of scheduling has to deal with work orders, part numbers, quantities, workstations, inventory (in-process, raw material, purchased items, and tools), personnel, and priorities. An important thing to keep in mind is that scheduling is a *component* of the process of executing the production plan and should be viewed accordingly: as a tool to *assist* in optimizing resource utilization measured against the specific goals of the management team. An on-line scheduling system within the manufacturing execution system that can absorb current event inputs, reevaluate on the basis of unplanned events, and communicate to other parts of the process can provide an enormous contribution toward optimization.

Modern quality assurance programs have a strong focus on the root cause of quality problems. It is not enough to repair a non-conforming item. The real issue is why the non-conforming event occurred. The same can be said of scheduling. If the developed schedule is the optimum use of resources, any deviation from that schedule will cause less than optimum results. If the root cause of schedule deviations could be recognized and resolved, it seems logical that the utilization of resources could be improved. For example, one manufacturer wanted to use finite scheduling, fore-casting to the minute when each work order or lot would arrive at a workstation during the next two weeks. Any event that did not occur within the scheduled time was examined to determine the cause of failure. Action was then taken to address the root cause, providing an excellent tool for continuous improvement.

References

American Production and Inventory Control Society, *APICS Dictionary,* eighth edition, APICS, Falls Church, VA, 1995.

AutoSimulations, 655 Medical Drive, Bountiful, UT 84010.

Berclain Group Incorporated, 3175, ch. Des Quatre-Bourgeois, Suite 100, Sainte-Foy (Quebec), G1W 2K7 Canada.

Goldratt, Eliyahu, *The Race,* North River Press, Croton-on-Hudson, NY, 1986.

Goldratt, Eliyahu, *The Haystack Syndrome,* North River Press, Great Barrington, MA, 1990.

Lawrence, K.D. and Stelios H. Zanakis, *Production Planning and Scheduling,* Industrial Engineering and Management Press, Norcross, GA, 1984.

Rembold, U., B.O. Nnaji, and A. Storr, *Computer Integrated Manufacturing and Engineering,* Addison-Wesley, Wokingham, England, 1993.

Vollman, T.E., William L. Berry, and D. Clay Whybark, *Manufacturing Planning and Control Systems,* Dow Jones-Irwin, Homewood, IL, 1984.

The Future for MES

MES is evolving to make systems easier to develop, design, install, and operate. New buzzwords are an unfortunate part of progress in systems, and MES is not without its share. The following are just a few of the newer terms heard today:

- Object-oriented programming
- OLE
- Client/server networks
- Global database/data warehouse/information storehouse
- Intranet
- Application programming interface
- Single-system image
- Plug and play

A major impediment to easier systems development and implementation is the difficulty in tying components such as MES support systems together. The computer industry is only now getting to the point where computers from different manufacturers with the same operating system can be connected. Soon it will be possible to combine different applications on different machines with different operating systems. A number of technologies look like they will offer real progress within the next few years, making the idea of *"plug and play"* (the ability to easily add or delete new software or equipment packages) a reality.

Client/server technology systems seem to offer a built-in technology to tie MES support functions to each other and/or to the core functions. Client/server technology is still developing and the definitions are quite varied. The following definitions are taken from *Essential Client/Server Survival Guide,* written by Robert Orfali, Dan Harkey, and Jeri Edwards, and published by Van Nostrand Reinhold (New York).

- **Service**—Client/server is primarily a relationship between processes running on separate machines. The server process is a provider of services. The client is a consumer of services. In essence, client/server provides a clean separation of function based on the idea of service.
- **Shared resources**—A server can service many clients at the same time and regulate their access to shared resources.
- **Asymmetrical protocols**—There is a many-to-one relationship between clients and server. Clients always *initiate* the dialog by requesting a service. Servers are passively waiting on requests from the clients.
- **Transparency of location**—The server is a process which can reside on the same machine as the client or on a different machine across a network. Client/server software usually masks the location of the server from the client by redirecting the service calls when needed. A program can be a client, a server, or both.
- **Mix and match**—The ideal client/server software is independent of hardware or operating system software platforms. It should be possible to mix and match client/server platforms.

The application ideas within the client/server approach would seem to allow a version of plug-in software application. An application example might be the decision to attach a time and attendance system package to the core MES. With the information requirements defined, the client (MES) requests a specific response (confirmation of skill for an employee) from the server (time and attendance). Without actually combining the systems, the information is requested and transferred. If this example is expanded to apply to all support functions or other MES systems on a corporate network, the possibilities are even more interesting.

Another major impact on future systems is *object-oriented technology*. This new approach to software programming will have a major impact on how information is moved within computer systems. The technology is not new. It was originally developed in Norway in the 1960s, but not until the past few years have applications been developed using this technology. Since then, many major software producers have object-oriented programming (OOP) in their newer products, including Windows 95 by Microsoft Corporation. Part of this technology is object linking and embedding (OLE), the Microsoft specification for application-to-application exchange.

Another term being heard more frequently is *application programming interface* (API). An API is the interface between a software systems application and the outside application. An example might be a time and attendance package with an API that can accept other MES package communications. The API is a logical interface, eliminating the need to go into the program itself.

The term *data warehouse* describes the concept of applying large amounts of data residing in interconnected databases to decision support systems (DSS). Manufacturing execution system sizes (measured in application breadth) are likely to grow substantially, generating huge amounts of data that should be accessible on-line. A revision to the data warehouse approach that will support on-line transaction processing (OLTP) systems is necessary and likely to evolve during the next few years. This methodology will take all information generated on the plant floor and place it in storage, available for immediate use through a company's worldwide system.

The world of manufacturing execution systems is quite young, and the application of true computer science to manufacturing questions is barely off the ground. Imagine where the airline industry would be without its application of computer science. How could worldwide banking and finance survive without these modern tools? Manufacturing applications have a much greater potential economic impact, and as more suppliers of systems to this market apply computer science to the real opportunities on the plant floor, the impact is likely to cause a snowball effect that will dramatically change how plants are managed. Manufacturing execution systems are a relatively new idea. As the saying goes, "You ain't seen nothin' yet."

References

Feldman, Len, *Windows NT,* Sams Publishing, Carmel, IN, 1993.

Inman, W.H., *Building the Data Warehouse,* John Wiley & Sons, New York, NY, 1992.

Orfali, Robert, Dan Harkey, and Jeri Edwards, *Essential Client/Server Survival Guide,* Van Nostrand Reinhold, New York, NY, 1994.

Taylor, David A., *Object-Oriented Technology: A Manager's Guide,* Addison-Wesley, New York, NY, 1990.

APPENDIX A

Application Examples

In this appendix, seven examples of manufacturing execution system projects are given to provide additional application background. These examples illustrate a variety of MES applications, ranging from space propulsion equipment manufacturing to dried fruit processing to biotech manufacturing. The companies highlighted are:

Caterpillar Inc.	York, Pennsylvania
Furnas Electric	Batavia, Illinois
Berlex Biosciences	Richmond, California
Mariani Packing	San Jose, California
United Technologies	San Jose, California
ITT Night Vision	Roanoke, Virginia
Covance Biotechnology Services	Research Triangle Park, North Carolina

Caterpillar Inc.
York, Pennsylvania
Oil Cooler Manufacturing Cell

During the past few years, Caterpillar has greatly intensified its commitment to improving manufacturing on a worldwide basis. Included in its program are Focused Factories and Factories With a Future. The Oil Cooler Cell is a Focused Factory plant within the York, Pennsylvania manufacturing facility which supplies oil cooling equipment for other Caterpillar divisions that resell the coolers as replacement parts or furnish them on Caterpillar products.

The MES was originally considered in 1990 with specific goals:

■ Reduce past due orders
■ Reduce inventory levels
■ Reduce lead times
■ Improve information flow
■ Eliminate repetitive tasks

The system has been in use since 1991 and has exceeded the original goals. When asked what the primary benefit of the system has been, the immediate answer is "*control.*" The information and the management of that information provide a proactive manufacturing execution system (MES) with always current knowledge and control of the manufacturing cell activities.

Oil coolers are made to order in lots of 1 to 500. There are 200 product part numbers. Input to the system, including work orders and shipping information, is received from the corporate computer center in Peoria, Illinois on a daily basis. Information output from the MES is generated every five minutes through polling by the York computer center for upload to Peoria. This provides the corporate computer center a current view of order status, labor reporting, equipment status, and shipping information.

The manufacturing cell is built around an automatic storage/retrieval system (AS/RS) that is used to store raw material and work-in-process material. The AS/RS has 500 storage locations with 48 interface points used for material delivery to 15 workstations.

The following terms are used within the facility and the MES:

■ **Workstation**—A place known to the MES where work is performed
■ **Interface Point**—AS/RS access point
■ **Inventory**—Fixtures and material used to build oil coolers
■ **User Interface**—Any operator or supervisor process

Work orders are received with part number, quantity, delivery date, and customer. The MES retrieves manufacturing plans, including the routing and bill of material for that part number, from the York computer center data library.

The system uses the work orders, manufacturing plans, and due date information to develop and provide a prioritized list of work. The cell supervisor uses the list to commit orders to be manufactured (staged). When the order has been staged, the system puts the work order in process by committing inventory and assigning work to workstations.

General System Functions

- **Planning System Interface**—Work orders can be downloaded to the MES from the corporate planning system each day. The MES is polled every five minutes for data regarding equipment status, material movement, and labor data. This polling is done by the computer center at York for upload to the corporate computer center in Peoria, Illinois.
- **Work Order Management**—Work orders are received from the corporate planning system daily. The supervisor has a number of actions to affect the order, including altering the planned schedule, staging orders, changing the status of an order, and assigning inventory.
- **Workstation Management**—The system uses routings for the part number identified in the work order to determine workstation assignment and loading.
- **Inventory Tracking and Management**—The MES maintains inventory data for all raw and work-in-process material and finished products in the Oil Cooler manufacturing cell.
- **Material Movement Management**—The AS/RS direction is a part of the MES system and is the primary method of material movement to and from workstations. Included in the system are movement algorithms designed to improve the rate of material delivery.
- **Data Collection**—Information on material receipts, material movement, labor, and equipment is collected and supplied to other system users.
- **Exception Management**—The system provides the supervisor with tools to respond to events that change the planned schedule.

The system provides screens that allow action or viewing by supervisors, workstation operators, and the dock station (receiving/shipping) operator. Supervisor screens are as follows:

1. **Actions**—Provides a supervisor with the same functions as the operator for the selected workstation.
2. **Amend AS/RS Priorities**—To amend the priority for any AS/RS interface point.
3. **Audit AS/RS Contents**—To list the containers located in all the AS/RS audit locations.
4. **Commit Current Schedule**—To commit to the schedule generated by the scheduling system and overwrite current cell and workstation order schedules.
5. **Display Log File**—To view errors and informational messages recorded by the system.
6. **Edit Committed Schedule**—To view the existing order schedule and amend or view order-related information.
7. **Equipment Reporting Status**—To report equipment breakdown or downtime for routine maintenance.
8. **Held Orders**—To view orders that have been held at workstations. The information displayed includes workstation, order details, reason entered by the operator, and time held.
9. **Initiate AS/RS Audit**—To initiate an audit of the AS/RS contents. The supervisor can audit a specified number of containers for a particular item or by random selection.
10. **Inventory Status**—To view the status of inventory within the cell.
11. **Local Applications**—To access company computer systems outside the MES.
12. **Material Receipt Rules**—To view and amend material receipt rules that govern where a container is stored.
13. **Retry Transport Missions**—To view and/or amend the current status of any outstanding material movement tasks.
14. **Transport Status**—To view the following information for all outstanding material movement tasks: container serial number, quantity in the container, move from location, move to location, move status, and the time the transport task was generated.

15. **View**—To view the status of all cell orders.
16. **Workstation Status**—To view the status of all workstations that are in progress.
17. **Execute Scheduling Module**—Directs the system to run the scheduling program.
18. **Amend Order Detail**—To amend certain details of a selected order.
19. **Amend Order Position (Cell Level)**—To resequence cell orders.
20. **Amend Order Position (Workstation Level)**—To resequence workstation orders.
21. **Amend Target Workstation**—To move orders from one workstation work list to another.
22. **Complete Transport Task**—To complete a failed transport task.
23. **Destage Material**—To release the material allocated to an order.
24. **Display Container Movement Details**—To display information on container movement in date/time order.
25. **Display Interface Point Availability**—To view information on interface point availability. For each interface point, the following information is displayed: time containers have been set at the interface point, time the interface point has been empty, total number of containers processed, and the arrival time of the current container.
26. **Hold Order**—To hold a cell order.
27. **Modify Container Details**—Allows a supervisor to change container detail information, including serial number, container type, location, item ID, and quantity.
28. **Move Container**—To move a container into, out of, or within the AS/RS.
29. **Release Held Order**—To release a held cell or workstation order.
30. **Retry Transport Mission**—To retry a failed transport task.
31. **Stage Material**—To stage material for an order.
32. **View Completed Orders**—To view all orders with a status of completed.

33. **View Order Details**—To view the order details.
34. **View Order Status**—To view the status information for all workstation operations for a specific order: workstation, reported quantity, estimated start, actual start, and status.
35. **View Manufacturing Plan**—To view the following manufacturing plan information for the selected order: operation plan, bill of material, and operation instructions.
36. **View Specific Workstation Orders**—To view all orders assigned to a specific workstation.

Each operator has a computer terminal at his/her workstation to provide communication with the system. Operator screens are as follows:

1. **Actions**—This screen allows an operator to view the staged orders for a workstation.
2. **Audit AS/RS Contents**—To list the containers located in all the AS/RS audit locations.
3. **Display Log File**—To display errors and informational messages recorded by the system.
4. **Equipment Reporting Status**—To report equipment breakdown or downtime for routine maintenance.
5. **Inventory Status**—To view the status of the inventory within the cell.
6. **Local Applications**—To access other applications within the computer network.
7. **Orders**—To view all orders for a workstation.
8. **Other Workstation Orders**—To view the status of another workstation's orders.
9. **Completed Orders**—To view completed orders.
10. **Transport Missions**—To view the following information for all outstanding AS/RS transport tasks: container serial number, quantity in the container, move from location, move to location, move status, and the time the task was placed.
11. **Deliver Empty Container**—To request the delivery of an empty container to a selected interface point.
12. **Deliver Item**—To view the items required for an order and then issue a request to deliver one or more items.

13. **Display Container Movement Details**—To display information on container movement in date/time order.
14. **Display Interface Point Availability**—To view information on interface point availability.
15. **Floor to Floor Move**—To inform the system that a container has been moved from a storage location to a workstation.
16. **Hold Order**—To hold a workstation order.
17. **Modify Container Details**—To change the details of a container.
18. **Move Container**—To move a container into, out of, or within the AS/RS.
19. **Release Held Order**—To release a held workstation order.
20. **Report Production**—To report information against a selected order: quantity produced, quantity scrapped, and order completion.
21. **Return Container**—To return containers to the AS/RS.
22. **View Manufacturing Plan**—To view the manufacturing information for the selected order.
23. **View Material Availability**—To view the availability of an item required for a particular order.
24. **View Order Details**—To view details of an order.
25. **View Order Status**—To view the status information for all operations for a specific order.

In addition to the screens provided for the supervisor and the operator, there are some screens unique to the dock station operator. These screens provide the following functions:

1. **Receive Raw Material**—To receive containers of raw material into the cell.
2. **Ship Finished Coolers**—To list all containers holding completed products and then select a number of containers to be shipped.
3. **Wanding Screen**—To enter container information into the system manually or automatically.

There are many MES systems within Caterpillar tied into a large network of data management systems. This system was installed by HK Systems of Milwaukee, Wisconsin.

Furnas Electric Company
Batavia, Illinois

Furnas Electric Company, a subsidiary of Siemans Energy & Automation, Inc., headquartered in Batavia, Illinois, produces electrical control devices and systems in five plants located in the United States, Canada, and Mexico. The company produces approximately 250 products, each of which has extensive variations and combinations, raising the number of unique products to more than one million. The products are distributed through various channels, including end users, distributors, and original equipment buyers that install Furnas equipment on their products. The products are essentially built to order, with order sizes ranging from 1 to 5,000. There is no typical order size. Accurate product assembly is a manufacturing requirement, as electrical control products carry a great liability when incorrect products are furnished.

Furnas began its consideration of an MES in 1992. The original objective was to provide work instructions at each workstation in order to eliminate the large amount of paper supplied with each shop order. There also was a strong desire for a method to make changes to orders previously released to the plant floor without reissuing the paperwork. The original goal of the system was to:

> ...develop a user friendly paperless manufacturing execution system that enables controlled electronic work order dispatch, provides access to engineering assembly instructions, enables quality detection and control, and provides production status information in support of the assembly cell, at a reasonable cost.

Production is done in assembly cells that are assigned a specific product or family of products. Each assembly cell is made up of a number of workstations, usually less than twenty. In some work cells, each workstations has a personal computer, but in other cells there is one personal computer for the entire cell. At present, there are four assembly cells on the MES, with more cells being added each year.

Planning-level functions are run on a mainframe computer using internally developed software located at corporate headquarters. Information is downloaded to the plant-level production planning system for the next day's production. Included in the download information are the work order number, part number, quantity, due date, and priority number. Any new work order or changes to existing work orders can be sent to the assembly cell by a planner at any time. This could simply be added to the current queue of work for that cell, or the planner might reprioritize the work orders in the queue. Orders are supplied to the workstation as sequenced by the planner.

The computer at the assembly cell displays only the work order with the highest priority. When the cell is prepared to begin work on the next work order, the file folder list indicating the information or material required for the selected order is displayed. The screen indicates the assembly drawings and the label information for a specific catalogue number. Through the CAD system (the product data library), the drawings for the part number specified on the work order are retrieved to provide complete assembly drawings with instructions for each workstation. The instructions and drawings are printed within the assembly cell and delivered to each workstation just prior to beginning work. The label information for the product to be furnished on this work order is then retrieved and printed at the assembly cell. Some assembly cells are directly connected to the quality assurance system. The MES system supplied by RWT Corporation, Mt. Prospect, Illinois, is designed with a feature that allows Furnas to create and add its own in-house applications. The MES runs on personal computers using Microsoft Windows NT operating system on a Novell network.

At this time, Furnas has met its original system goals, eliminating most of the plant floor paperwork, using on-line capabilities to effect changes in released orders, and improving quality through the immediately available and always current work instructions. The company gives a lot of credit to the MES for being able to reduce its work order cycle time from two weeks to two days, improve product and process quality, reduce indirect labor, and provide better customer service.

Berlex Biosciences*
Richmond, California

Within pharmaceutical and biotechnology companies, the focus of MES shifts from resource tracking to a heavy emphasis on Electronic Batch Records (EBR). The EBR focus is the result of the vast amounts of paperwork required to document a compliant manufacturing process, as well as the extensive time and effort required to document and approve the batch records prior to shipping the final product. All aspects of operating in a Current Good Manufacturing Practices (CGMP) regulated environment, such as operator training records, equipment calibration and maintenance records, material lots used, in-process testing steps and results, etc., must be recorded and documented.

With the maturation of the biotechnology industry, areas outside drug discovery and clinical development become more important and face greater scrutiny from economic and regulatory points of view. Production of material for clinical studies has to happen in a timely manner, be fully CGMP compliant, and eventually be scalable to a commercial level with as few process changes as possible.

The application of MES in this environment is the same as in other industries—to link the activities within manufacturing, such as Material Requirements Planning (MRP), Laboratory Information and Management System (LIMS), process control, etc. Information from these planning and control systems flows to the MES, and in turn these systems are updated with real-time plant floor data. MES provides instant access to plant floor activity to include the status of work orders, material lots and location, labor, equipment, and facilities. This information can be accessed quickly so that decisions can be made with accurate and timely information. The significant benefits of MES include:

- CGMP compliance (the system will enforce and document compliance with approved specifications)
- Reduction of human error

---

* Furnished by Base 10 Systems. This application example is taken in part from an article published in *Biopharm,* 10(3), 46–48, March 1997, an Advanstar publication.

- Material tracking and verification, ensuring that only the correct materials and lots are used in the manufacturing process and that the expiration date has not passed
- Through the use of electronic work instructions, the system ensures manufacturing operations are completed in compliance with approved specifications
- Equipment and facility monitoring to ensure that equipment is in the proper stage for production (i.e., maintenance, validation, calibration, and cleaning have been performed as required)
- Monitoring operator training records to confirm the operator is current with the latest versions of appropriate manufacturing documents
- Security that provides complete control over who can access the system
- Immediate access to process-related documents such as safety information
- Costs tracking and analysis

The MES reduces paperwork, human error, and the time required to rearview and approve batch records, while expediting the manufacturing processes. All data associated with a batch record can be consolidated in the system. It is no longer necessary to compile paperwork from numerous departments and perform tedious reviews. In one central location, all batch information is assembled simply by reviewing the activities that were deviations from the approved specifications, along with the associated corrective actions.

The process currently in use at Berlex Bioscience where the MES has been applied includes the following stages:

- Inoculum preparation (includes maintaining the Master Cell Bank and Manufacturers Working Cell Bank and growing the reactor seeds in T-flasks and roller bottles)
- Media and buffer preparation, either automated or manual, with powders in mixing tanks
- Fermentation, adherent or suspended mammalian cells in perfusion reactors

- Purification, consisting of several chromatographic and filter steps
- Freeze drying
- Fill finish, sealing and packaging of the vials with the product

Each of these steps has a different level of automation, ranging from completely manual (e.g., inoculum preparation) to highly automated (e.g., freeze drying). For each batch, daily production activities occur, and information is recorded relative to the activities being performed. Due to the varying degrees of automation, either manual, electronic, or combined data entry occurs. It is important to be able to access the results easily and quickly, either during or after a batch is manufactured.

Two important aspects of the protein production process should be emphasized. One is the backward and forward traceability of lots that constitute a final product batch. The process starts with one or several vials and proceeds via several T-flasks and roller bottles where the inoculum is grown up into several reactors. Those are harvested on a continuous basis, where identifiable lots are associated with the harvest period. Sometimes reactors are split— using one reactor to inoculate a second one—constituting new lots, although lots originating from a "mother" reactor are then combined for column runs in the purification suite, sometimes using different numbers at each of the purification steps. Identifying the history of a particular lot can therefore be a major undertaking, requiring access to dozens of log sheets. Having all information in an easily accessible database removes a great burden and also potentially shortens times during inspection by the regulatory authorities.

The other aspect is the close interaction between the production activities and the LIMS utilized by Quality Control (QC). All materials that enter production must be qualified (with the results and release information provided by LIMS), and a constant stream of samples has to be taken at all process steps. Alerts for sampling, labeling sample containers, requesting analysis by QC, and proper handling of the test results are normal production activities processed through the LIMS.

Much of the justification for the system centers around **inventory control**, management of material, and improving operational efficiency. All material movement is tracked, from receipt of materials to shipment of fulfilled orders. The screen for the first process to be supported, receipt of raw materials, is shown in Figure 1. All manufacturing items are entered into the system, with many requiring full QC testing. The MES provides the link between the material movement and the laboratory. This screen shows how information

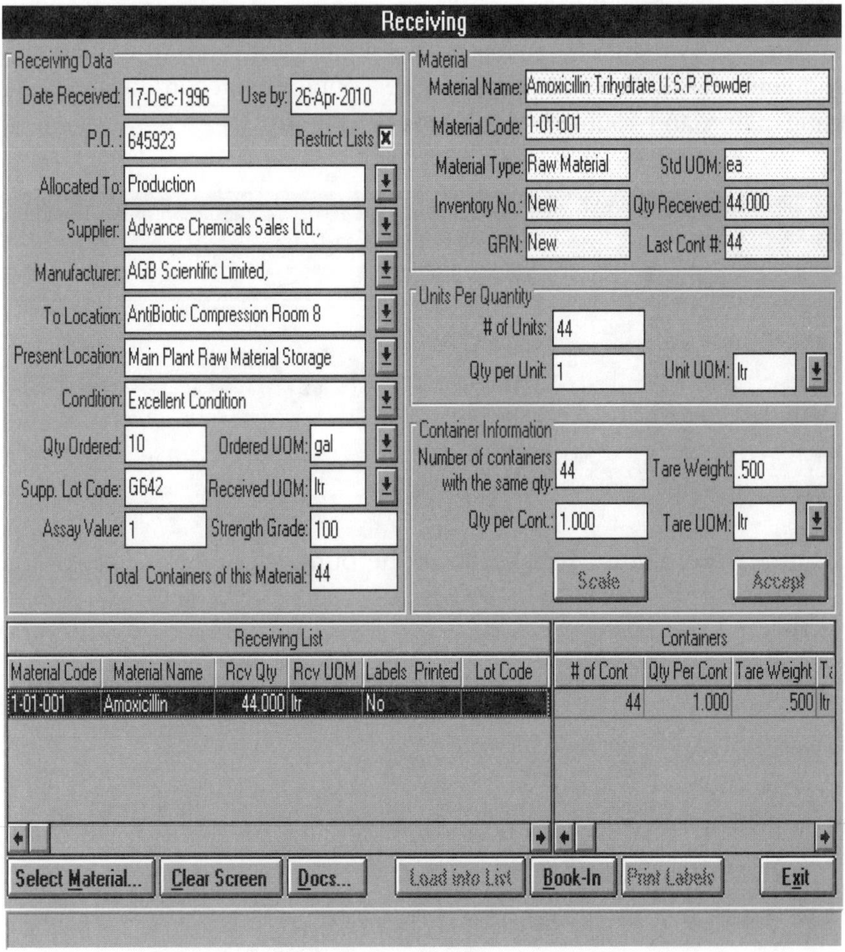

Figure 1   Receiving Screen

is entered to the system during material receiving, allowing checks on the following items:

■ That the material was ordered
■ That the correct quantity was delivered
■ That the material passes visual inspection

The MES automatically notifies the QC laboratory that material is in quarantine awaiting release. Some items may be released directly to stock, either because they come from approved vendors or because they do not require full QC testing.

The system generates bar-coded labels for both the material containers as well as for the sample containers according to sampling plans and regulatory guidelines (Figure 2).

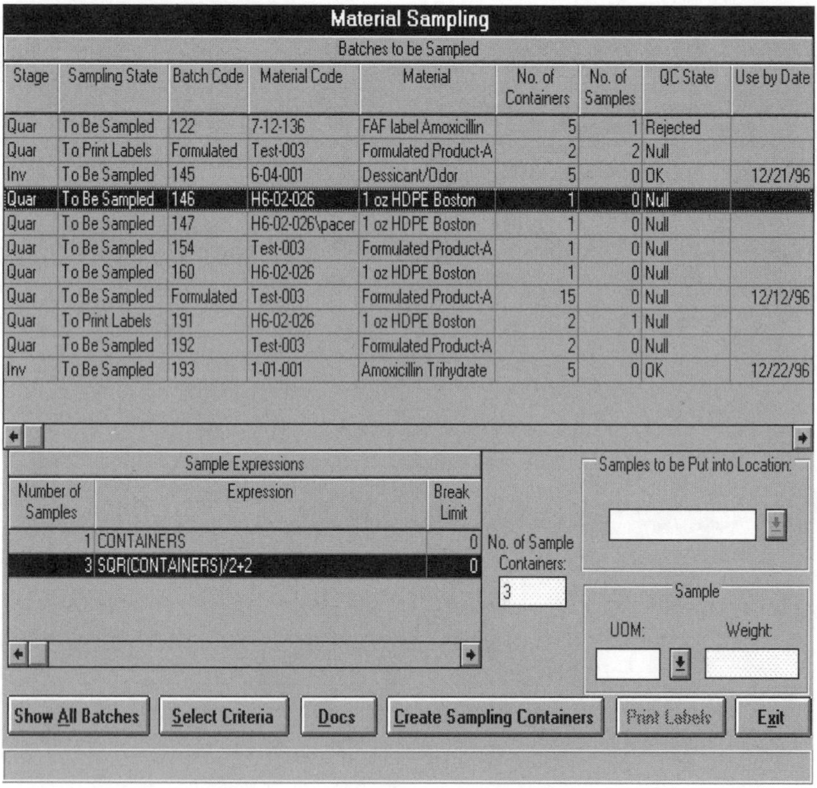

Material Sampling

Batches to be Sampled

Stage	Sampling State	Batch Code	Material Code	Material	No. of Containers	No. of Samples	QC State	Use by Date
Quar	To Be Sampled	122	7-12-136	FAF label Amoxicillin	5	1	Rejected	
Quar	To Print Labels	Formulated	Test-003	Formulated Product-A	2	2	Null	
Inv	To Be Sampled	145	6-04-001	Dessicant/Odor	5	0	OK	12/21/96
Quar	To Be Sampled	146	H6-02-026	1 oz HDPE Boston	1	0	Null	
Quar	To Be Sampled	147	H6-02-026\pacer	1 oz HDPE Boston	1	0	Null	
Quar	To Be Sampled	154	Test-003	Formulated Product-A	1	0	Null	
Quar	To Be Sampled	160	H6-02-026	1 oz HDPE Boston	1	0	Null	
Quar	To Be Sampled	Formulated	Test-003	Formulated Product-A	15	0	Null	12/12/96
Quar	To Print Labels	191	H6-02-026	1 oz HDPE Boston	2	1	Null	
Quar	To Be Sampled	192	Test-003	Formulated Product-A	2	0	Null	
Inv	To Be Sampled	193	1-01-001	Amoxicillin Trihydrate	5	0	OK	12/22/96

Sample Expressions

Number of Samples	Expression	Break Limit
1	CONTAINERS	0
3	SQR(CONTAINERS)/2+2	0

No. of Sample Containers: 3

Samples to be Put into Location:

Sample UOM: Weight:

Show All Batches | Select Criteria | Docs | Create Sampling Containers | Print Labels | Exit

Figure 2   Material Sampling Screen

Once laboratory testing has been carried out and the disposition of the sample has been determined, the inventory can be released to stock from this screen. In some cases, partial release is required if some containers do not comply with QC standards. The MES is designed to respond to this, as well as to reject an entire delivery. For each of the possible outcomes from the QC manager, the MES generates the appropriate bar-coded label (Figure 3).

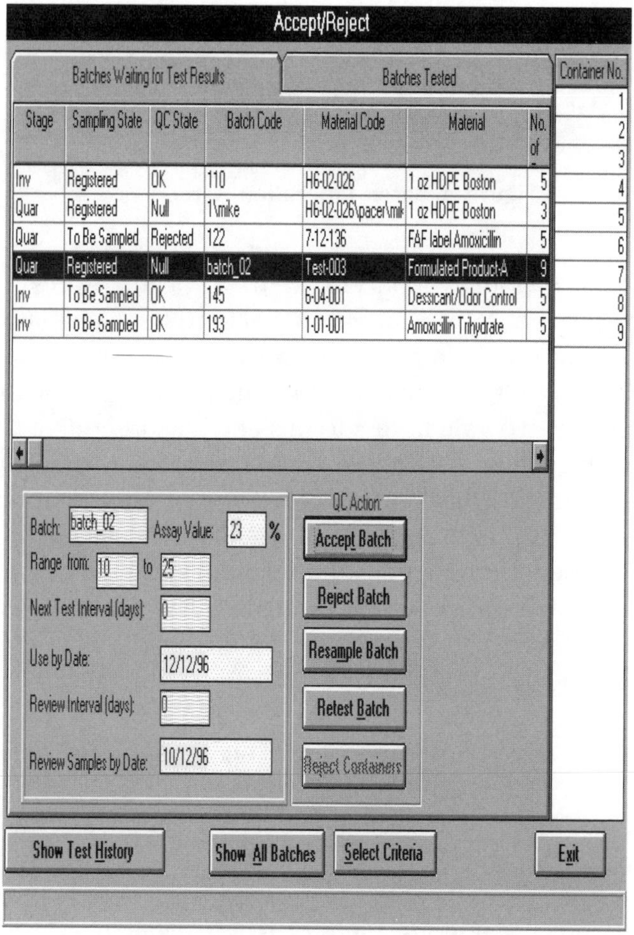

Figure 3   Accept/Reject Screen

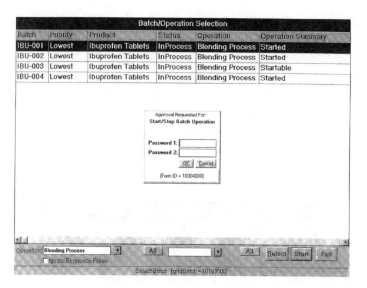

Figure 4 Batch Operation Screen

Security is a primary concern in this system. Access to the functionality of the MES is through a control center, shown in Figure 4. Each of the main functional areas is accessed through the click of a button, with context-sensitive help displayed at the bottom of the screen. To gain access to this and all subsequent screens, an electronic signature is required. In FDA-regulated industries, an electronic signature is the key to acceptance of MES in controlling batch processing to the standards of CGMP. The system supports a "two-token" identification, consistent with FDA guidance on specification and control of electronic signature. The two tokens are a bar-coded employee badge (physical token) and a four-digit PIN number (known token). Regulatory guidelines can call for an even higher degree of security, such as a retina scan or fingerprint pattern, for some applications. These higher security levels are usually required for "open systems," where more general access is allowed. Since MES is classified as a "closed system" by the FDA, restricted to use by employees and their agents, two-token control is sufficient.

Access to unit operations for processing material is through a prioritized list of outstanding batches. Access to each batch is controlled through the electronic signature. In this screen, the two tokens are requested as "password one" and "password two," and a clear indication is given in red that the signature is required to "start/stop batch operation." Such clear indication of the reason for requesting the signature is one of the requirements of the FDA. Once the material for processing is selected, the work instruction can be presented. Any deviations from approved processes can be recorded and approved according to privilege level.

The introduction of MES helps ensure regulatory compliance, reduce manual labor, and provide immediate access to information for all production stages. It also provides connectivity between the different activities within the enterprise and leads toward a fully integrated and automated operation.

References

Black, R.S., Control-Based MES for the Pharmaceutical, Biotechnology, and Fine Chemical Industries, *Pharmaceutical Technology,* 11(19), pp. 38–48, 1995.

Gold, D.H., Validation: Why, What, When, How Much, PDA *Journal of Pharmaceutical Science and Technology,* 50(1), pp. 55–60, 1996.

Ostrovski, S. and G.C. du Moulin, Achieving Compliance in Validation and Metrology for Cell and Gene Therapy, *BioPharm,* 8(9), pp. 20–28, 1995.

PDA Letter, 32(3), pp. 17, 27–28, 1996.

Roth, G., J.M. Kubiak, J.F. Long, and G.M. Schoofs, An Automated System to Produce Cell Culture Media from Liquid Medium Concentrates, *BioPharm,* 8(9), pp. 31–45, 1995.

Mariani Packing Company Incorporated
San Jose, California

Mariani Packing is a processor and packer of dried fruit products, including raisins, prunes, apricots, peaches, pears, etc. The primary production facility is located in San Jose, California, with three fruit-supplying plants outside of the metropolitan area. The company employs approximately 350 people and ships over 1,500 different products worldwide through three market channels (retail packaged products, export products, and fruit ingredients) to food processors such as cereal manufacturers.

The MES has three primary uses: (1) track and monitor inventory lots from initial receipt through processing, (2) collect and monitor data from numerous processing (manufacturing) points, and (3) provide an archival capability for manufacturing information.

The MES system focuses on the lots of fruit received by the company. A lot can be bins or truckloads, measured in pounds. When the fruit is received, a lot number is assigned, and information is collected and assigned to the lot number. The information includes fruit size, variety, the crop year, lot size, lot number, etc.

Processing the fruit follows specific paths. Each path processing point is identified in the logic of the MES. The result, in the simplest terms, is that a lot is moved from location A to location B with defined actions performed on the lot at each point. Additionally, the system allows the operator to take actions regarding each lot, such as splitting a lot into smaller lots, combining lots into larger lots, and assigning lots for shipment.

An important aspect for this business is the yield of each lot of fruit and the costs associated with each lot. Much of the data developed and collected by the system is used to measure the current yield and then archived to build an information base for future analysis and comparison.

The system presents the main menu for each operator after logging on. After choosing one of the initial menu functions, the lower level screens are presented for operator action. The various screens offer the following functionalities.

WIP Tracking System*
Mariani Initial Menu

1. **Start**—To initiate the formation of a lot and assign the lot number

---

* Supplied by Camstar Systems, Inc., Campbell, California.

2. **Split**—To split a lot into two or more smaller lots
3. **Multiple Lot Combine**—To combine two or more lots into one larger lot
4. **Super Move**—To choose an alternate lot routing and an alternate set of data collection items
5. **Assign Lots to Inventory or Ship**—To collect certain lots into a group
6. **Group Shipment**—To assign ship to locations for a group of lots
7. **Receipt**—To indicate material has arrived at the ship to location
8. **Group Move**—To move a group of lots
9. **WIP Tracking Menu**—To provide multiple on-line views of work-in-process
10. **Assign P1 Number and Dry Bin Tares**—To assign tare data to a bin
11. **Lot Status by Operation**—To provide lot status information by operation
12. **View Lot Inquiry**—To provide access to information about a specific lot
13. **Lot Status by Grower Number**—To provide status information by lot and grower number
14. **Lot Status by Lot Number**—To provide status information on each lot
15. **Lot Status by Lot Group**—To provide status information on each lot group
16. **Mariani Reports Menu**—To provide a listing of possible reports, allowing the operator to choose the desired report
17. **Download Inventory Data to MAPICS System**—To transfer inventory information to the office accounting system
18. **Group Receive**—To indicate material (a lot or group of lots) has been received
19. **Actual Ship Date Entry**—To enter the actual date the truck will leave the plant and includes a group ID number
20. **Lot History/Move Summary**—To list the historical information for a specific lot

21. **Lot Status by Product Group**—To list the current status of lots within a product group
22. **Available Inventory Query—All Plants**—To provide information on the current inventory within the system
23. **Print Shipping Manifest in San Jose**—To print a shipping manifest
24. **Print Shipping Manifest in Marysville**—To print a shipping manifest

The main menu screens call lower level screens to receive input to the system or to display requested information. Mariani also does extensive data collection from various process points for quality measurement. Information is keyed into the MES or collected automatically and then analyzed and presented through the MES or by a third-party quality assurance system.

The system as provided by the vendor is made up of the following modules and tailored to fit the customer:

Tracking and Monitoring

- **Process Model**—Builds a model of the manufacturing process
- **Work-in-Process Tracking**—Tracks work-in-process by lot, batch, or unit serial number
- **Material Control**—Defines all product relationships, including bills of material, options, alternate process paths, and product grades
- **Real-Time Production Monitor**—Provides on-line snapshots of current throughput and yield results
- **Unit-Level Tracking**—Tracks work-in-process by unit or serial number within a lot
- **Activity-Based Costing**—Provides full activity-based costing of work-in-process

Quality Management

- **Quality Data Collection**—Collects and edits any user-defined quality, test, or parametric data from the process

- **Data Extract and Analysis**—Provides a function for extracting and correlating all parametric, yield, and quality data collected in a relational database format
- **Statistical Process Control (SPC)**—Monitors and controls SPC parameters for out-of-control conditions
- **Specifications/Document Control**—Provides password and sign-off control of specifications, instructions, and process data tolerances

Planning and Scheduling

- **Lot Scheduling**—Provides real-time forward or backward scheduling of in-process lots
- **Dispatching**—Provides on-line display of operator dispatch lists at each workstation
- **Support System Interface**—Provides an interface to allow external scheduling systems of supply dispatch lists

System Interfaces

- **Planning-Level Interface**—Provides standard interface formats and programs for connecting to planning systems
- **Automation Module**—Provides interfaces and formats to allow cell controllers, workstations, and automated equipment to exchange data with the MES

The following additional capabilities are also part of the system:

- **Security**—Security is controlled separately in each plant environment. A single point of entry is used to maintain security profiles in the plant. Security passwords are encrypted to prevent unauthorized access.
- **Archive Control**—There is a full archive control subsystem to automate the off-line storage and retrieval of critical lot traceability and process data.
- **Tailorability**—The software is designed to be tailored to the user's environment without requiring changes to source programs. This is done through the use of user-defined tables and codes and soft-coded application terminology.

United Technologies Corporation, Chemical Systems Division* San Jose, California

Chemical Systems Division manufactures solid propellants, rockets, and advanced propulsion systems for space and missile systems such as Titan, IUS, Trident, Tomahawk missiles, and the space shuttle. The MES serves three manufacturing areas:

- **Avionics**—Highly reliable electronic/electromechanical components that are part of control systems for a variety of rocket motors.
- **Small Motor Processing**—Fabrication and final assembly of rocket motors used to launch airborne vehicles, separate solid rocket boosters, or act as igniters.
- **Large Motor Processing**—Fabrication and final assembly of rockets to support earth orbit and intercontinental distances. Includes inert motor preparation, propellant mix and cast, and final fabrication and assembly work centers.

The original objectives of the system were developed through a fifteen-member team:

- Reduce lead times.
- Automate the handling of work instructions and paperwork.
- Automate the collection and reporting of product quality data from the shop floor.
- Establish a real-time link between planning functions and production.
- Collect labor and non-labor cost data at the operation, lot, and work order or job level.
- Integrate the hourly time card system with the labor data collection system.

---

* The MES system and most of the information for this case study were provided by RWT Corporation of Mount Prospect, Illinois. Additional assistance in reviewing the system with the author was provided by Chuck Aiello and Millie LaFave of Chemical Systems Division.

- Facilitate easy-access two-way communication between machine tool operators and support functions.
- Provide production management with timely, accurate information regarding the product flow through the manufacturing process.

The resulting system uses MRPII data for plant floor execution. The MRP system has been in place since 1996. Bi-directional data transfer from the MRPII system to the MES interface module provides work order information, bills of material, routings, quantities, due dates, production status, work center status, and labor and material reporting.

With each work order there is an extensive electronic work instruction that details each operation to be performed. The electronic work instruction for the designated part number is retrieved from the document control computer and added to the work order. Once the work instruction is attached, the system has access to electronically stored engineering drawings, set-up sketches, tool drawings, and engineering information. Quality-related required data items that need to be captured as part of the production process are identified or preset and assembled in the electronic work instruction folder. Examples of required data items include part serial number, component lot or component serial number, tool number, tool calibration date, torque values, operator certification level, etc.

Once the folder has been prepared, production control uses the order dispatch capabilities to sequence work to be performed. This releases the work order and associated folder to the production floor. The operator views the dispatch, selects the appropriate work order, and begins to log time and activities to the order. After collecting production results and labor data, the information can be used for timekeeping, further labor analysis, and cost analysis.

Since each part produced must be traced through its complete genealogy, the operator uses the folder to tell what required data items must be captured. If required, the operator indicates which part serial number is being worked on and what components and component serial numbers are being used. If special tools are necessary, they can be verified in the system, and torque values are captured. If out-of-range conditions exist, the system flags the operator, and a non-conformance report needs to be generated.

When an engineering change notice occurs, an allied configuration control system electronically notifies the planners and operators, and a work order can be immediately put on hold. The planner can identify the changes and update the electronic work instruction folder so work can continue.

All of the related as-built information entered into the MES is tracked during the production process. This provides Chemical Systems Division's customers with a complete electronic record of every operator, every activity, and every component used in the production process—thus facilitating the electronic sell-off.

The system follows the function outline of MES.

Core Functions

- **Planning System Interface**—There is a direct link from the MRPII system to the MES that provides bi-directional data transfer of work orders and work order information. This link is under revision to accommodate the installation of a new enterprise planning system.
- **Work Order Management**—Work orders are received into the MES and sequenced through production.
- **Workstation Management**—Work is sequenced into workstations by the systems and the workstations are connected to the system for information display and data collection.
- **Inventory/Material Management**—The system tracks the work-in-process between workstations and provides information regarding tools and components.
- **Data Collection**—Extensive information developed throughout the production process is collected and distributed to users and the work order files.

Support Functions

- **Documentation Control**—The management of drawings, instructions, and other information necessary for manufacturing is available through the PC DOCS document management system.

- **Quality Assurance**—Quality assurance requirements and the electronic work instruction folder. Information is collected at the workstation through the user terminal during the production process.
- **Time and Attendance**—Labor information is collected through the terminal at the workstation.
- **Genealogy**—The system collects information pertinent to a specific part or work order to provide a complete history of the production process.

This system, as do most MES systems, provides many functions that help users in ways additional to production management. Some examples are as follows:

- **Memo**—The "memo main menu" selection provides the capability to send a note to the operator or any individual signing on to the workstation.
- **Audit Trail**—The "audit trail" selection provides the capability to view the transactions recorded on the transaction log. The following information is displayed:
  - □ **Date**—The date and time of entry of the transaction into the log
  - □ **Origin**—The point of origin of the transaction
  - □ **Type**—The type of transaction
  - □ **Param**—The subcategory of the transaction type
  - □ **Data**—Other information associated with the transaction

Other examples can include security, error alarms, error logs, error recovery, system query language, and database maintenance.

The system at Chemical Systems Division has eliminated the significant amount of paperwork previously used in production, improving the information provided to the plant floor and the data collected from production. With less paper to handle, and less chance of providing the wrong document, lead times are also being reduced. Future plans include implementing a finite capacity scheduling system to improve the sequencing of work within each business unit.

ITT Night Vision*
Roanoke, Virginia

ITT Night Vision (ITT NV) is the leading manufacturer and supplier of night vision equipment for the military, law enforcement, and commercial markets. The plant, located in Roanoke, Virginia, employs 650 people and has annual sales of more than $100 million. ITT NV is a business unit of ITT Defense & Electronics. Its manufacturing process has more than 400 separate steps using 200 different chemicals in working vacuums. The process employs molecular-level manufacturing tolerances and a full spectrum of scientific disciplines.

In 1990, management had a vision: a new information system that would support ITT NV's quest to remain competitive, introduce new products to new markets, and achieve world-class manufacturing status. The old "home-grown" system with islands of automation had to go. A new system that could turn data into information to run the business was needed.

Management formed a feasibility study team. The team was cross-functional, with members from all areas of the company, but it was clear the project would be manufacturing driven. The team had a mission to research, justify, and recommend to management a system to meet the needs of the business today and grow with the company for the future.

Defining the needs of the business and the end users was the first task. Product cost needed to be lowered by finding ways to improve yields and reduce scrap. Reduced time to market for new products and reduced cycle times were also key elements. In addition, the system had to be user friendly and flexible to a continuous process improvement environment.

The current systems were studied to identify weaknesses, and the market was researched to identify available packages. Several point solution packages were ruled out quickly because they could

---

* This case study information was prepared by Peggy Taylor, PROMIS Project Manager, ITT Night Vision. The system was furnished by Promis Systems Corp. of Nashua, New Hampshire.

not meet all the needs of the company. The team identified an MES as the right path to investigate. Two MES suppliers were selected for the in-depth study, and the team began the evaluation process. Members of the team demanded hands-on experience with the packages and tried out the software on the company process. Using these models (specific to ITT NV's process), the team conducted on-site demonstrations for the management and manufacturing staff. The team made many reference calls to users of both MES systems and made a series of site visits to facilities using these systems.

The team concluded that both MES systems could do the job and meet the basic requirements of the end user. The final decision was based on the following key elements:

- Cost
- Performance
- Support
- Multiple process versions active
- Flexibility to high-volume process changes

The team forecasted the system would drive an improvement in product yield and used this as justification, placing the return on investment at 2.2 years. The system was purchased in the fourth quarter of 1992.

ITT NV trained an initial development/implementation team of sixteen people. It was clear the project would be manufacturing developed, and the company built the team with engineers, supervisors, operators, and data coordinators. The company rented a building near the plant to isolate the development team and let them focus on the project. Team members dedicated 60% or more of their day to the project while maintaining their regular jobs, often working twelve- to fourteen-hour days. Initially the team planned a "slam-dunk approach," which was a grand idea to set up the whole factory on the MES at once. However, defense cuts and labor reductions drained resources. The company downsized and a new company president came on board. With the trained team cut in half and the project behind schedule, the team reevaluated the approach.

Time for a reality check! How would the team accomplish the task with reduced resources? The team selected the smallest core of the factory where the "home-grown" system could be dismantled while keeping the remaining areas running on the old system. The smallest core of the factory turned out to be 50% and covered six major production areas and two product lines.

Training was recognized as a major task. The team had to train 300 people on three shifts and had two weeks to do it. They felt the biggest obstacle to overcome was the fear of change. The operators needed to learn how the MES worked without worrying about how it would fit into their normal jobs. A three-step approach was developed:

- **MES Overview**—Required training for the entire plant. Concentration was on the MES concept, terminology, and communication of project goals.

- **Generic Tracking Model**—A unique and creative approach was taken using a Lego helicopter model kit. A simulated helicopter assembly line was set up, with all products and parts tracked by operators on the MES. Each step of the model taught a different function of the MES. It took the fear out of learning the MES. Even the company president put together a helicopter and tracked with the system.

- **Specialized Training**—Taught the specific use of the MES for night vision manufacturing, specific to individual production areas and individual jobs.

The system was functional, and most problems involved the introduction of the work-in-process inventory in midstream. Around-the-clock coverage with constant communication addressed the problems. Elimination of paper in half the areas within the first week gave team members a confidence level about the capabilities of the system they had delivered to the factory. Despite the obstacles, the team had developed and implemented the system in half the factory in seven months. The team refined the initial implementation and started the rollout for the remaining areas of the factory, one by one.

The system provides the following benefits:

- Yield improvement has exceeded the predicted 2%, with an actual improvement of 5% that continues to improve.
- Turnaround time to introduce a database for new products to manufacturing reduced from six months to six weeks.
- Reduced some rework categories from as high as 25% to as low as 2%.
- Rework loops have been defined and documented, and accountability has been established.
- Improved the inventory control by 5% over three years, with 1995 being the best physical inventory in division history.
- Reduced cycle times.
- Improved data input accuracy and automatic process calculation.
- Improved process control with complete traceability from start to finish.
- On-the-spot data availability to end users.
- One common database for manufacturing information decisions.

The system functions primarily as a data collection system through more than 300 operator terminals at workstations and more than 400 process steps. Production scheduling information is developed by production planning and the planning system (run once per week) and is manually entered into the MES. Inventory management and material movement are accomplished outside the MES. Some information is transferred from the statistical process control system to the MES. Through the presentation of information screens at each workstation, the operator can be given instructions and/or given an outline of the data to be input by responding to prompts within the MES. Exceptions are readily dealt with, and a complete tracing record of each manufactured item is available for review.

This is an extensive system in a very complicated manufacturing facility that begins as a complex chemical process and results in a discrete item, a night vision viewer. The MES has produced significant results and will grow through system changes as the business makes even better use of the information.

Covance Biotechnology Services*
Research Triangle Park, North Carolina

Covance Biotechnology Services is the country's largest full-service recombinant outsource manufacturing facility. It is located within the established pharmaceutical and rapidly growing biotechnology community of Research Triangle Park, North Carolina. The state-of-the-art, multimillion-dollar plant is capable of producing bulk therapeutic proteins and peptides from both mammalian and microbial expression systems under CGMP compliance.

Manufacturing areas are dedicated to a single product during execution of the inoculum, recovery, and purification processes. In manufacturing areas with closed systems, such as fermenters and bioreactors, simultaneous production of multiple products is possible. Two purification suites are available for efficiently handling both small- and large-scale purification processes.

Covance is a representative example of an FDA-regulated manufacturing facility. At the manufacturing execution system level, it is characterized by the following key attributes:

- Recipes must be managed at different levels of detail.
- Recipes contain raw material, products, and by-products, and a variance is expected between the planned versus actual production in all cases where the recipe is not followed exactly.
- A large number of measurements must be made on a continuous basis (e.g., quantities, flow rates, ranges, process conditions) rather than in discrete terms (e.g., each, counts).

---

* Material for this application example was taken from *A MES System for Regulated Process Manufacturing* by S. Zafar Kamal, Ph.D., of ABB Industrial Systems. Contributors to the paper include: Mark Witcher, Ph.D., Covance; Dan Costa, Covance; Najeeb Anwer, ABB Industrial Systems; David Foster, ABB Industrial Systems; Scott Mohrmann, ABB Industrial Systems; Rajesh Ramachandar, ABB Industrial Systems; Robert Russell, ABB Industrial Systems; David Tatich, ABB Industrial Systems; and Sai Yesantharao, ABB Industrial Systems. The system was designed by Covance and ABB. It was developed and integrated by ABB Industrial Systems.

Subsequently, there are extensive requirements on data capture and reporting.

■ The emphasis is on operator-initiated/monitored control rather than high-speed autonomous control.

The MES system described here provides the information management, monitoring, and reporting capabilities to the Covance manufacturing environment. This has been accomplished by the design, implementation, and integration of systems that satisfied the functional requirements for the different business functions related to the manufacturing process.

The manufacturing system at Covance is schematically shown in Figure 5. Other corporate and/or manufacturing support systems (e.g., financial accounting, maintenance) are not shown.

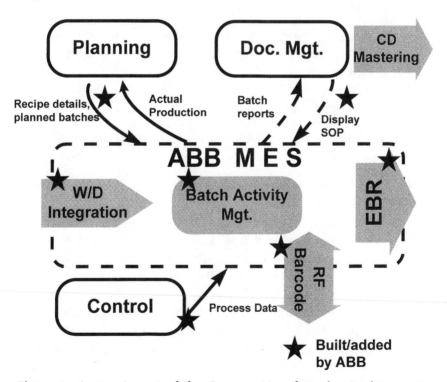

Figure 5   System Layout of the Covance Manufacturing Environment

The systems enclosed in clear boxes are purchased off-the-shelf applications, and those shown in shaded boxes were developed by ABB Industrial Systems. In addition to the developed applications, ABB provided the "software" necessary to integrate the systems identified in Figure 5.

The selection of the systems identified in Figure 5 was based on providing the best-in-class functionality to achieve the manufacturing business objectives at Covance. Some of the key features of the overall system are listed as follows:

- Separation of the planning and execution functions
- Creation of single/unique master data sources (e.g., item master, warehouse master, formula/recipe definition)
- A consolidated production/process information data repository for after-the-fact reporting and analysis
- Maintaining single data entry points for each manufacturing function
- Minimizing the use of paper at the plant floor, resulting in on-line collection of all data from manual as well as automated activities
- Operator guidance and verification of actions, including electronic signatures

The following MES functions are provided by the current Covance system:

1. **Planning System Interface**—The interface between planning and the MES provides a "seamless," configurable mechanism for the transfer of information between the two environments. Specifically, it makes the production demand visible to the MES and in turn enables the MES to report actual production (material consumption and production) to the planning system.
2. **Inventory Tracking and Management**—All materials used in the production process are appropriately tracked and accounted for. These materials include direct raw material, expendable materials, work-in-process, and reusable and bulk-manufactured materials. Materials are controlled and tracked

at two levels: through the planning system and through the MES system at the plant floor. The MES system material tracking and control is accomplished through the Weigh & Dispense system and the Batch Activity Management (BAM) system.

3. **Workstation Management**—Covance's manufacturing process requires that a large number of transactions be initiated and captured at the plant floor. Transactions encompass various manufacturing activities (e.g., initiating material flow, sampling) which may be prescribed or scheduled. Furthermore, transactions involve data entry by operators, data capture from process control, and the printing of reports and labels. Transactions are accomplished through minimal keyboard entry using hand-held radio frequency terminals and various bar-code-reading data entry devices.

4. **Work Order Management through Production Operations Guidance, Validation, and Monitoring**—All operations conducted on the plant floor are monitored with respect to:
   □ **Ownership**—To track who did what
   □ **Time**—The time of the transaction
   □ **Relevance**—Verification that the transaction is relevant to the process steps
   □ **Accuracy**—The ability to view and report expected versus actual values, with appropriate comments for exceptions
   □ **Lineology**—The interrelationship of transactions, material movements, etc. to previous and subsequent operations.

5. **Reporting—Electronic Batch Records and CD-ROM Archiving**—Detailed information related to production operations, material movement, process conditions, etc. is captured and reported on demand. Process data, instructions, and other production details associated with the manufacture of a "batch" of a product are captured and stored in a relational database. This enables electronic storage and retrieval of Electronic Batch Records (EBR) and archiving of these batch records to an indexed, self-contained CD-ROM.

6. **Production Operations Information Capture, Control, and Automation**—Wherever possible, manufacturing processes are

automated using the ABB Advant process control system. The control system and the manufacturing operations are coordinated to satisfy the data capture, validation, and tracking requirements. Information is captured through direct data entry by the operator using radio frequency terminal bar code readers or by accessing the control system database.

7. **Document Management**—The ability to manage (create, control, track, update, view, circulate, electronic signatures, provide security for, and distribute) large quantities and types of documents within the corporation and to external entities (e.g., customers, regulatory agencies, vendors) is provided through a Document Management System (DMS). The MES environment updates the DMS with production reports and provides structured access to key documents needed by the operators.

The following operational areas within Covance use the MES:

- Quality Control and Quality Assurance (including the lab)
- Process Operations and Engineering
- Document Control
- Purchasing, Receiving, and Warehouse Operations

The automation, monitoring, and data acquisition functionality that connects the manufacturing devices and instrumentation to the MES is accomplished through the Open Control System (OCS), which is capable of complete plant-wide automation. The architecture of the system accommodates functional and geographical distribution of the hardware, software, and database over the plant while allowing system-wide access to distributed data. It utilizes a modular architecture to permit a wide range of system configurations and facilitate system flexibility and expandability.

The OCS has a global, relational real-time database with a real-time relational database management system that is distributed among the system units. During system operation, each system unit has its portion of the global database only, but is also able to access data from anywhere in the system. The control system software provides the following database management functions:

- Direct access of I/O
- Data security
- Real-time access in a distributed environment
- Validity check on operator and programming language commands
- Query capability (search and retrieval based on conditional requests) of the real-time database, including engineering and historical data

In addition to the process control system, the MES environment utilizes some features of the system for batch control, monitoring, and data acquisition. The features relevant to the MES system are as follows:

- Conforms to Instrument Society of America (ISA) SP88 batch standards and guidelines
- Low-level detailed (automation) recipe creation and management
- Allows for the identification of raw materials by lot, amount, and physical location; material characteristics may also be assigned to each lot
- Batch equipment specification and allocation
- Batch history capture in SQL-accessible form to enable acquisition of all historical data

The DMS is capable of collecting, managing, displaying, and archiving documents in the MES environment. The system also provides security features, including the capability to record electronic signatures for identification. These documents may originate from different sources (e.g., Oracle Report Writer, Microsoft Office applications, Graphics). The built-in workflow management capability within the DMS is used to direct document approval and distribution.

The DMS controls:

- Storage and retrieval
- Workflow-based circulation and routing
- User access control and security

- Distribution
- Printing of controlled documents

The DMS package makes it possible to view current copies of documents (e.g., material safety data sheets and standard operating procedures) from all Windows-based computers at the facility (including the Advant operator stations).

The BAM system guides the operator through various sequential steps involved in the manufacturing process. It replaces paper forms with electronic data entry screens, bar code scanner, and direct data import from the control system. The BAM provides reliable data integrity and validation of information collected.

The BAM concept involves the generation of specific transactions for the collection and display of process and event data. Each transaction utilizes forms for configuration and subsequent operator entry of data. Transactions permit manual data entry, process control data acquisition, acknowledgment of operations, security checks, etc. Transactions are analogous to the steps in a procedure or the pages of a paper document (e.g., batch records, instruction sheets). Information collected with these transactions is stored in a relational database to be made available for assembly into an EBR document.

At present, the system allows more than thirty different transaction types covering operations in a pharmaceutical manufacturing facility. After each transaction or as a part of each transaction, the operator may be prompted to provide a comment. These comments can be of the following types:

- A **general observation report** is used to document information that does not influence the product quality or manufacturing procedure.
- A **deviation report** is used to document non-conforming events or data. The input of a deviation report will automatically enforce the input of a second supervisor's comment.

During the configuration phase, the transactions are configured in the correct sequence with the manufacturing product and equipment information and are assembled into an executable manufacturing procedure.

In addition to event tracking for batch manufacturing, the system allows for the following:

- Back/forward linking of transactions, to enable setting up dependencies between transactions previously executed or to be executed in the future
- Maintaining lot identity among repackaged (as dispensed) containers

All data collected with the transactions is stored in a relational database for access by other systems and eventually into an EBR. The key modules of the BAM system are described as follows:

- **Detailed Recipe Configuration**—Batch Activity Management Configuration is an "authoring" system which lets product managers/quality control personnel create/modify/release manufacturing recipes—called "manufacturing procedures." The BAM system allows for the definition of manufacturing instructions as a combination of appropriate manual steps, operator-initiated events, data capture commands to the OCS, and operator-initiated control commands to the OCS environment. Once configured, these steps are used to guide the operator through sequential steps in the manufacturing process.
- **Recipe Execution**—Batch Activity Management Execution (BAME) process is invoked by an operator when performing plant floor manufacturing operations. Execution is typically done from a hand-held device that displays detailed instructions based on the manufacturing instruction set defined in the configuration phase. The system continually provides instructions to the operator and solicits operator input. The BAME has user validation capability at the field level to validate all user inputs and events.
- **Material Reconciliation and Reporting**—The BAME environment maintains details on lot/batch relationships with re-

spect to containers and material usage at specific manufacturing operations. Material consumption and production are reconciled before reporting to the planning system.

APPENDIX B

Sources of MES
Information and Vendors

Associations

APICS
American Production and Inventory Control Society
500 West Annandale Road
Falls Church, Virginia 22046-4274
Telephone: (703) 237-8344
Sponsor of annual APICS conference and trade show

Institute of Industrial Engineers (IIE)
25 Technology Park
Norcross, Georgia 30092
Telephone: (770) 449-0461

ISA
Instrumentation Society of America
P.O. Box 12277
Research Triangle Park, North Carolina 27709
Telephone: (919) 549-8411
Sponsor of annual ISA conference and show

MESA International
Manufacturing Execution Systems Association
303 Freeport Road
Pittsburgh, Pennsylvania 15215
Telephone: (412) 781-9511
Sponsor of conferences on MES

Society of Manufacturing Engineers (SME)
One SME Drive
P.O. Box 930
Dearborn, Michigan 48121-0930
Telephone: (313) 271-1500
Sponsor of annual Autofact conference and trade show

Magazines

Managing Automation
Thomas Publishing Company
5 Penn Plaza
New York, New York 10001
Telephone: (212) 629-1551

Manufacturing Systems
191 South Gary Avenue
Carol Stream, Illinois 60188
Telephone: (708) 665-1000

Performance Advantage
Lionheart Publishing, Inc.
Suite 299, Cumberlund Parkway
Atlanta, Georgia 30339

Directories

Buyers' Guide and Resource Directory
MESA International
303 Freeport Road
Pittsburgh, Pennsylvania 15215
Telephone: (412) 781-9511

MES Directory and Comparison Guide
Thomas Software Directories
Thomas Publishing Company
5 Penn Plaza
New York, New York 10001
Telephone: (212) 629-1112

Trade Shows and Conferences that Include MES

MES Roundtable Conference
Sponsored by MESA International
303 Freeport Road
Pittsburgh, Pennsylvania 15215
Telephone: (412) 781-9511
Bi-annual conferences held in various locations

Autofact
Sponsored by Society of Manufacturing Engineers (SME)
One SME Drive
P.O. Box 930
Dearborn, Michigan 48121-0930
Telephone: (313) 271-1500
Held in the fall in Detroit or Chicago

APICS Conference
Sponsored by APICS (American Production and Inventory
Control Society)
500 West Annandale Road
Falls Church, Virginia 22046-4274
Telephone: (703) 237-8344
Held in the fall in various locations

International Industrial Engineering Conference
Sponsored by Institute of Industrial Engineers (IIE)
25 Technology Park
Norcross, Georgia 30092
Telephone: (770) 449-0461

National Manufacturing Week
Sponsored by Reed Exhibition Companies
P.O. Box 7247-8458
Philadelphia, Pennsylvania 19170
Telephone: (203) 840-5878
Held in Chicago in March

ISA Conference
Sponsored by Instrumentation Society of America
P.O. Box 12277
Research Triangle Park, North Carolina 27700
Telephone: (919) 549-8411
Annual conference held in various locations

There are MES suppliers in many other trade shows specific to a given industry.

APPENDIX C

Glossary*

This terms defined here are frequently used in discussions relating to manufacturing computer applications including MES.

Acceptance Sampling—The process of sampling a portion of goods for inspection rather than examining the entire lot. The entire lot may be accepted or rejected based on the sample even though the specific units in the lot are better or worse than the sample.

Activity-Based Costing—A cost accounting system that accumulates costs based on activities performed and then uses cost drivers to allocate these costs to products or other bases such as customers or markets. It is an attempt to allocate costs on a more realistic basis than direct labor or machine hours.

Ad Hoc Sequencing—A sequencing method in which work orders are ordered according to an operator-entered specification.

Aggregate Inventory—The inventory for any grouping of items or products involving multiple stockkeeping units.

Agile Manufacturing—This can mean different things, but the general idea is the ability to respond very quickly to changes in what

---

* Many of the definitions listed here are reprinted with permission of APICS—The Educational Society for Resource Management, Falls Church, Virginia. The selected material comes from the *APICS Dictionary,* seventh edition (1992) and *APICS Dictionary,* eighth edition (1995).

is being manufactured or what is to be manufactured. Examples might include manufacturing to individual unit orders (such as mail-order computer manufacturing) or automobile manufacturing where the lot size is usually one. One definition might be delivering the right product at the right time, the right cost, and the right quality with unplanned, unpredictable, continuously changing demand.

AGVS—Abbreviation for Automatic Guided Vehicle System.

Alarm Summary—A log of system alarm conditions. Entries are made into the alarm summary as alarm conditions occur.

Allocated Item—In an MRP system, an item for which a picking order has been released to the stockroom but the item has not yet been sent from the stockroom.

Alternate Routing—A routing alternative to the primary or preferred routing.

American Standard Code for Information Exchange (ASCII)—A standard seven-bit character code used by computer manufacturers to represent 128 characters for information interchange among information processing equipment.

APICS—Acronym for the American Production and Inventory Control Society.

Application Software—A computer program or collection of programs that performs a designated function.

Artificial Intelligence (AI)—Computer programs that can learn and reason in a manner similar to humans. Sometimes called expert systems.

AS/RS—Acronym for Automatic Storage/Retrieval System, a rack storage system with automatic loading (storage) and unloading (retrieval).

Assemble-to-Order—An environment where a product or service can be assembled after receipt of the customer's order.

Assembly Parts List—A list of all parts that make up a particular assembly.

Available Inventory—The on-hand inventory balance minus allocations, reservations, back orders, and quantities held for quality problems.

Backflush—The deduction from inventory records based on the count of items produced.

Backward Scheduling—A technique for calculating operation start dates and due dates. The schedule is computed starting with the due date for the order and working backward to determine the required start date and/or due dates for each operation.

Bar Code—A series of bars and spaces representing encoded information that can be read by electronic readers.

Bill of Material (BOM)—An ordered list of the parts, subassemblies, assemblies, and raw material that define a product.

Bill of Resources—A listing of the required capacity and key resources needed to manufacture one unit of a selected item or family.

Bottleneck—A production activity where actual capacity is less than the demand.

Bucket Scheduling—Frequently used to describe time periods used to develop time-phased schedules.

Buffer—A quantity of materials awaiting further processing. It can refer to raw materials, semifinished stores or hold points, or a work backlog that is purposely held behind a workstation. In the theory of constraints, buffers can be time or material and support throughput and/or due date performance.

Buffer Management—In the theory of constraints, a process in which all expediting in a shop is driven by what is scheduled to be in the buffers. By expediting this material into the buffers, the system helps avoid idleness at the constraint and missed due dates.

Business Process Reengineering—A procedure that involves the fundamental rethinking and radical redesign of business processes.

CAD—Acronym for computer-aided design, the use of computers in engineering drawing and design.

CAM—Acronym for computer-aided manufacturing, the use of computers to program and control production equipment.

Capacity Requirements Planning (CRP)—A methodology for determining the labor and machinery capacity required to accomplish production based on the open orders and planned orders in the MRP system as the production requirement.

CAPP—Acronym for computer-aided process planning.

Cellular Manufacturing—See *Work Cell.*

CGMP—Acronym for Current Good Manufacturing Practices, a term often used in pharmaceutical manufacturing or other manufacturing where compliance with government regulations and procedures is necessary.

CIM—Acronym for computer-integrated manufacturing, a philosophy of using computers to accomplish total integration of the manufacturing organization into a coherent integrated whole.

COMMS—Acronym for Customer-Oriented Manufacturing Management System.

Configuration—The arrangement of components as specified to produce an assembly.

Constraint—Any element or factor that prevents a system from achieving a higher level of performance with respect to its goal. Constraints can be physical, such as a machine or lack of material, but they can also be managerial, such as a policy or procedure.

Constraints Management—An approach to scheduling using constraint theory to optimize production results.

Continuous Process Improvement—A managerial method and philosophy of continuous small-step improvements in production.

Critical Path Method (CPM)—A method of project planning and scheduling using activities and associated times.

Critical Ratio—A sequencing rule based on the ratio of time remaining ÷ work remaining.

$$\frac{\text{Due date} - \text{Now}}{\text{Lead time remaining}}$$

A resulting ratio of less than one indicates the job is behind schedule. A ratio of more than one indicates the job is ahead of schedule.

Critical Ratio Sequencing—A sequencing method in which work orders are ordered according to their critical ratios.

Cycle Counting—An inventory auditing method where inventory is counted on a cyclic schedule rather than once a year.

Cycle Time—In materials management, this refers to the length of time from when material enters a production facility until it exits.

Database—A computer file processing approach designed to manage and index computer data items.

Data Collection—The process of recording a transaction and transmitting it to a computer.

Data Library—A location for storing information relating to the manufacturing process.

Decision Support Systems (DSS)—Computer systems that are used to analyze data and evaluate courses of action by creating logical data reports.

Delivery Lead Time—The time between the receipt of an order and the delivery of the product.

Demand—A need for a particular product or component.

Demand Rate—A statement of requirements in terms of quantity per unit of time.

Detailed Scheduling—The assignment of starting and/or completion dates to operations to show what must be done if a work order is to be completed on time.

Discrete Manufacturing—Production of distinct items such as automobiles, appliances, or computers.

Dispatching Rule—The logic used to assign work order priorities at a workstation. *Syn.*: sequencing rule.

Download—The transfer of information from a computer to another computer lower in the system hierarchy.

Employee Empowerment—The practice of giving non-managerial employees the responsibility and the power to make decisions regarding their jobs or tasks.

Engineering Change Order—A revision to a design released by engineering to modify a part.

Enterprise Resources Planning (ERP)—A later evolution of manufacturing planning systems usually including distribution, product data management, and supplier management. ERP systems are built around later information technology including database management systems, client/server computer systems, and improved communication capabilities between systems such as CAD product data libraries and plant floor data collection devices.

Ergonomics—The approach to job design that focuses on the interactions between the human operator and the environment.

Exception Management—The response to unplanned events that cause the planned production schedule to change.

Expert System—A type of artificial intelligence computer system that mimics human action by using rules and heuristics.

Facilities—The physical plant and equipment.

Fault Tolerance—The ability of a system to avoid or minimize the disruptive effects of defects by utilizing some form of redundancy or extra design margins.

Final Assembly—The highest level assembled product, as it is shipped to a customer.

Final Assembly Schedule—A schedule of end items to finish a product in a make-to-order environment.

Finite Forward Scheduling—A scheduling technique that builds a schedule by proceeding sequentially from the initial period to the final period.

Finite Loading or Scheduling—A concept of scheduling by simulating work order starting and stopping at each workstation based on the finite capacity limit of each workstation.

First Come/First Served—A sequencing rule based on the time of arrival at the workstation.

Flexible Capacity—The ability to operate manufacturing equipment at different production rates by varying staffing levels and operating hours or starting or stopping at will.

Float—The amount of work-in-process inventory between two manufacturing operations.

Forecast Horizon—The period of time into the future for which a forecast is made.

Forward Scheduling—A scheduling technique that begins with the known start date and computes the ending or finish date.

Functional Design Specification—A detailed description of a proposed computer system detailing data input and output points, man–machine interface definitions with screen layouts, and all expert or exception rules.

Gantt Charts—Named after Henry Gantt, these charts describe the use of bar charts to present scheduling information.

Group Technology—A method of part classification based on the physical similarity of parts.

Indented Bill of Material—A form of multilevel bill of material. It exhibits the highest level parents closest to the left margin, and all of the components going into those parents are shown indented to the right of the margin. All subsequent levels of components are indented further to the right.

Infinite Loading—Calculation of the capacity required at workstations regardless of the capacity available to perform the work.

Input/Output Devices—Modems, terminals, or other devices whose designed purpose is to provide data to and from a computer system.

Inventory—Those stocks or items used to support production (raw materials and work-in-process items), supporting activities (maintenance supplies), and customer service (finished goods and spare parts). In the context of MES, inventory might include any item required to accomplish production, such as material, tools, fixtures, drawings, instructions, etc.

Inventory Control—The activities and techniques of maintaining the desired levels of items in inventory.

Inventory Shrinkage—Losses of inventory resulting from scrap, deterioration, pilferage, etc.

Inventory Turnover—The number of times that inventory cycles, or "turns over," during the year.

Islands of Automation—Stand-alone pockets of automation (robots, AS/RS, a CAD/CAM system) that are not connected into a cohesive system.

Just-In-Time (JIT)—A philosophy of managing inventories to the lowest level possible while maintaining production. Material arrives just in time before running out.

Lead Time—The span of time required to perform a process or series of operations.

Level Schedule—A schedule that has distributions of material requirements and labor requirements that are as even as possible.

Limiting Operation—The operation with the least capacity in a series of operations with no alternate routings. The capacity of the total system can be no greater than the limiting operation.

Line Balancing—The balancing of the assignment of the elemental tasks of an assembly line to workstations to minimize the number of workstations and to minimize the amount of idle time at all stations for a given output level.

Local Area Network (LAN)—A high-speed data communication system for linking electronic data-handling and -producing devices.

Lot Number Traceability—Tracking parts by lot numbers to a group of items to aid in later identification and usage tracking.

Lot Operation Cycle Time—The length of time required from the start of set-up to the end of cleanup for a production lot at a given operation, including set-up, production, and cleanup.

Lot Size—The amount of a particular item that is ordered or manufactured.

Lot Sizing—The process of determining lot size.

Lot or Order Splitting—Dividing a lot into two or more sublots and simultaneously processing each sublot.

Lot Traceability—The ability to identify the lot or batch numbers of consumption and/or composition for manufactured items.

Machine Loading—The accumulation by workstation of the hours generated from the scheduling of operations for released orders by time period.

Manufacturing Automation Protocol (MAP)—An application-specific protocol based on the International Standards Organization's open systems interconnection. It is designed to allow communication between computers from different vendors in the manufacturing shop floor environment.

Manufacturing Control System (MCS)—A term sometimes used to describe the planning system layer.

Manufacturing Instruction (MI)—A set of detailed instructions for carrying out a manufacturing process.

Manufacturing Resources Planning (MRPII)—A system of planning and scheduling resources within a manufacturing facility.

Material Requirements Planning (MRP)—A planning tool used to establish manufacturing schedules and loading. Beginning with product quantities required in the master production schedule, MRP calculates the quantity for each line item (part number) in the bill of material. After the quantity of each item required is determined and measured against inventory on hand, the quantity to make or purchase is calculated. The element of lead time is then taken into the equation to determine when these items must be ordered or manufactured.

MESA International—An association of suppliers of manufacturing execution systems.

Min-Max System—An order point replenishment system where the "min" is the order point and the "max" is the "order-up-to" inventory level.

Model—A representation of a process or system that attempts to relate the most important variables in the system in such a way that analysis of the model leads to insights into the system.

Mortgaged Material—Material that has been reserved for a specific order or orders.

Move Ticket—A document used in dispatching to authorize and/or record movement of a job from one work center to another.

Multilevel Bill of Material—A display of all the components directly or indirectly used in a parent, together with the quantity required for each component.

Need Date—The date when an item is required for its intended use.

Net Change MRP—An approach in which the material requirements plan is continually retained in the computer. Whenever a change is needed in requirements, a partial explosion and netting is made for only those parts affected by the change.

Netting—The process of calculating net requirements.

Network—The interconnection of computers, terminals, and communications channels to facilitate file and peripheral device sharing as well as effective data communications.

Non-Significant Part Numbers—Part numbers that are assigned to each part but do not convey any information about the part. They are identifiers, not descriptors.

Open Systems Interconnection (OSI)—A communication system where a user can communicate with another user without being constrained by a particular manufacturer's equipment.

Operation—An identifiable processing step to produce an item.

Operation Description—The details or description of an activity or operation to be performed. This is normally contained in the routing document and could include set-up instructions, operating instructions, and required product specifications.

Operation Sequencing—A simulation technique for short-term planning of actual jobs to be run at each workstation based upon capacity, priority, existing work force, and machine availability.

Operations Sequence—The sequential steps for an item to follow in its flow through each workstation. This information is frequently called the routing.

Operation Start Date—The date when an operation should be started in order for the order to meet its due date.

Order Point System—Inventory method that places an order for a lot whenever the quantity on hand is reduced to a predetermined level known as the order point.

Overlapped Schedule—A manufacturing schedule that "overlaps" successive operations. Overlapping occurs when the completed portion of an order at one workstation is processed at one or more succeeding workstations before the pieces left behind are finished at the preceding workstation.

Parallel Conversion—A method of system implementation in which the operation of the new system overlaps with the operation of the system being replaced.

Pilot Test—In computer systems, final acceptance testing of a new business system using a subset of data with engineered test cases and documented results.

Planning Horizon—The amount of time a schedule extends into the future.

Planning System/Planning Layer—The term used to describe the accumulated computer functions found in MRP, MRPII, and ERP systems that determine (among other things) what is to be manufactured.

Point Reporting—The recording and reporting of milestone manufacturing events.

Priority—In a general sense, the relative importance of jobs (i.e., the sequence in which jobs should be worked on).

Process Flow Scheduling (PFS)—A generalized method for planning equipment usage and material requirements that uses the process structure to guide scheduling calculations. PFS is used in flow environments common in process industries.

Process Manufacturing—Production that adds value by mixing, separating, forming, and/or performing chemical reactions. It may be done in either batch or continuous mode.

Process Steps—The operations or stages within the manufacturing cycle required to transform components into intermediates or finished goods.

Product Data Management (PDM)—Computer systems used to manage and archive product information such as drawings, bills of material, routings, process information, assembly illustrations, etc.

Product Genealogy—A record of the history of a product from its introduction into the production process through its termination.

Productive Capacity—The additional output capabilities of a resource when operated at 100% utilization.

Product Structure—The sequence that components follow during their manufacture into a product. A typical product structure would show raw material converted into fabricated components, components put together to make subassemblies, subassemblies going into assemblies, etc.

Programmable Logic Controller (PLC)—An electronic device that is programmed to test the state of input process data and to set output lines in accordance with the input state. Programmable controllers provide factory floor operations with the ability to monitor and rapidly control hundreds of parameters, such as temperatures, pressures, etc.

Pull System—The production of items only as demanded for use or to replace those taken for use.

Push System—The production of items at times required by a given schedule planned in advance.

Quality—Conformance to requirements or fitness for use.

Quality at the Source—A producer's responsibility to provide 100% acceptable quality material to the consumer of the material.

Quarantine—The setting aside of items from availability for use until all required quality tests have been performed and conformance certified.

Queue—The jobs at a given workstation waiting to be processed.

Queue Time—The amount of time a job waits at a workstation before set-up or work is performed on the job.

Rate-Based Scheduling—A method for scheduling and producing on a periodic rate.

Raw Material—Purchased items or extracted materials that are converted into components and/or products.

Real Time—A technique of coordinating computer system data processing with external related physical events as they occur, thereby permitting reporting of conditions promptly.

Redundancy—A backup capability, coming from either extra machines or extra components within a machine, to reduce the effects of breakdowns.

Repetitive Manufacturing—A form of manufacturing where various items with similar routings are made across the same process whenever production occurs.

Replenishment Lead Time—The total period of time that elapses from the moment it is determined that a product is to be reordered until the product is back on the shelf available for use.

Request for Proposal (RFP)—A document that describes the requirements for a system or product and requests proposals from suppliers.

Requirements Definition—In MES implementation, a process or document that defines the current problems and opportunities, the objectives and goals of the system, and the expected results of the project.

Resource—Anything that adds value to a product or service in its creation, production, and delivery.

Response Time—In computer system terminology, response time is the elapsed time between the initiation of a transaction and the results of the transaction. It is often used to describe the time a computer screen takes to display a requested result.

Routing—The list of sequential operations or steps detailing the method of manufacture of an item.

Run Time—Operation set-up and processing time.

Scheduling Rules—Basic rules that can be used consistently in a scheduling system.

Sequencing Rules—Sequencing rules are used to determine which job to run next at a workstation.

Set-up Time—The time required for a specific machine, resource, workstation, or line to convert from the production of the last good piece of lot A to the first good piece of lot B.

Shop Floor Control—Generally used to describe the process of tracking, dispatching, and scheduling work orders through manufacturing.

Shop Packet—A package of documents used to plan and control the shop floor movement of an order. The packet may include a work order, operation sheets, engineering information, move tickets, time tickets, etc.

Shortest Processing Time (SPT)—A sequencing rule that prioritizes jobs in ascending order by processing time.

Simulation—The technique of using representative or artificial data to reproduce in a model various conditions that are likely to occur in actual performance of a system and make "what if" evaluations.

Slack Time—The difference in calendar time between the scheduled due data for a job and the estimated completion date. If a job is to be completed ahead of schedule, it is said to have slack time; if it is likely to be completed behind schedule, it is said to have negative slack time.

Split Lot or Order—A manufacturing order quantity that has been divided into two or more smaller quantities, usually after the order has been released.

Standard Time—The length of time that should be required to (1) set up a given machine or operation and (2) run one part/assembly/batch/end product through that operation.

Statistical Process Control (SPC)—Monitoring a process by analyzing outputs using statistical techniques that provide feedback to be used in maintaining or improving process capability.

Statistical Quality Control (SQC)—The use of statistical techniques in the quality function.

Structured Query Language (SQL)—A computer language that is a relational model database language. It is English-like and non-procedural and provides the ability to define tables, screen layouts, and indexes.

Synchronized Production—A manufacturing management philosophy that includes a consistent set of principles, procedures, and techniques where every action is evaluated in terms of the global goal of the system. Both kanban, which is part of the JIT philosophy, and drum-buffer-rope, which is part of the theory of constraints philosophy, represent synchronized production control approaches.

Theory of Constraints (TOC)—A management philosophy developed by Eliyahu M. Goldratt which is useful in identifying core problems of an organization.

Time and Attendance—A collection of data relating to an employee's record of absences and hours worked.

Time Phasing—A technique for expressing future demand, supply, and inventories by time period.

Time Stamping—Tracking the time of occurrence with each transaction.

Time Standard—The predetermined times allowed for the performance of a specific job or operation.

Traceability—The registering and tracking of parts, processes, and materials used in production by lot or serial number.

Value Added—The actual increase in utility gained when a product is transformed from raw material to finished inventory.

Visual Control—The control of authorized levels of inventory in a way that is instantly and visibly obvious.

Wait Time—The time a job remains at a workstation after an operation has been performed until it is moved to the next operation.

Wand—A device connected to a bar code reader to identify a bar code, or the process of using a wand to scan a bar code.

"What If" Analysis—The process of evaluating alternate strategies by answering the consequences of changes to forecasts, manufacturing plans, inventory levels, etc.

Where-Used-List—A listing of every parent item that calls for a given component, and the respective quantity required, from a bill of material file.

Windows NT—A computer operating system product of Microsoft Corporation with 32-bit addressing and multitasking capabilities.

Work Cell—A group of workstations usually having some common purpose or focus.

Work-in-Process (WIP)—A product or products in various stages of completion throughout the plant.

Work Order—An order to manufacturing authorizing products to be made.

Workstation—A physical location where work is accomplished.

Yield—The ratio of usable output from a process to its input.

Index

variance, 35
Laboratory Information and Management System (LIMS), 124
LANs, see Local area networks
Large motor processing, 136
Lawrence, K.D., 109
Lead time, for part number, 8
Least set-up sequencing rule, 102
Least work remaining sequencing rule, 101
Licensing information, 94
LIMS, see Laboratory Information and Management System
Line balancing, 101
Local area networks (LANs), 67
Longacre, Andrew, 81
Lot traceability, 35

M

Mainframe computer, 122
Management, Systems and Synthesis, 19
Managing Software Development Projects, 90
Man–machine interface, 90
Manual movement, 34
Manufacturers Working Cell Bank, 125
Manufacturing
  Automation Protocol (MAP), 14
  cycle time, 2
  environments batch process, 23
    continuous process, 23
    defined routing variations, 23
    discrete part, 23
    in-line variations, 23
    MES in different, 21–24
    repetitive discrete part, 23
  information, archival capability for, 132
  plan, execution of, 1
  Resources Planning (MRPII), 7, 8
Manufacturing execution systems (MES), 1–6
  core functions, 5, 25–40
    data collection, 5, 25, 35–36

data library, 37–38
exception management, 5, 25, 36–37
inventory tracking and management, 5, 25, 32–33
material movement management, 5, 25, 34
planning system interface, 5, 25, 26–27
scheduling, 37
system of integrated functions, 38–40
work order management, 5, 25, 27–29
workstation management, 5, 25, 29–31
definition, 2
equipment used in, 63–81
  computer hardware, 63–69
  data entry, 69–80
  voice recognition, 78
future for, 111–114
implementation
  data collection/acquisition, 4
  dispatching production units, 4
  document control, 4
  labor management, 4
  maintenance management, 4
  operations/detail scheduling, 4
  performance analysis, 4
  process management, 4
  product tracking and genealogy, 4
  quality management, 4
  resource allocation and status, 3
information and vendors, sources of, 153–156
needs of
  plan and execute, 24
  plan and have ready, 24
  reporting, 24
proactive, 3
selection team, 84
Software Evaluation/Selection, 87
support functions, 5–6, 41–55
  documentation/product data